Dorsi Germann / Eberhard Gohl

Participatory Impact Monitoring

Booklet 1:
Group-Based Impact Monitoring

A Publication of Deutsches Zentrum für Entwicklungstechnologien – GATE
A Division of the Deutsche Gesellschaft für Technische Zusammenarbeit (GTZ) GmbH

The authors:

Dorsi Germann, sociologist and graphic artist, spent four years working in a community development project in Senegal. For the last fourteen years, she has been a consultant on adult education, appropriate technologies, technics of communication and visualization, project management, monitoring and evaluation, organizational development and participatory methods in Africa, Asia and Latin America, mainly working for GTZ and FAKT.

Eberhard Gohl, economist and sociologist, spent a few years in Turkey, Peru and Bolivia. For eight years, he worked mainly with FAKT, DSE-ZEL and GTZ as a consultant for project management and organisation development. At present, he works as Controller in the German Protestant Churches' funding NGO „Bread for the World".

Die Deutsche Bibliothek – CIP-Einheitsaufnahme

Participatory impact monitoring : a publication of Deutsches
Zentrum für Entwicklungstechnologien – GATE, a division of
the Deutsche Gesellschaft für Technische Zusammenarbeit
(GTZ) GmbH. – Braunschweig ; Wiesbaden : Vieweg.
  ISBN 3-528-02086-5
NE: Deutsches Zentrum für Entwicklungstechnologien <Eschborn>

Booklet 1. Group based impact monitoring / Dorsi
  Germann/Eberhard Gohl. – 1996
NE: Germann, Dorsi

The author's opinion does not necessarily represent the view of the publisher.

All rights reserved
© Deutsche Gesellschaft für Technische Zusammenarbeit (GTZ) GmbH, Eschborn 1996

Published by Friedr. Vieweg & Sohn Verlagsgesellschaft mbH, Braunschweig/Wiesbaden

Vieweg is a subsidiary company of the Bertelsmann Professional Information.

Printed in Germany by Lengericher Handelsdruckerei, Lengerich

ISBN 3-528-02086-5

CONTENTS

Do you need a new tool to manage your project? 3

1. What does monitoring mean? 4

    How difficult is monitoring? 6

    What should be monitored? 7

    How is it done? 8

2. Steps in introducing and carrying out
group-based impact monitoring 9

    Preliminary step: What is known about the context? 11

    Step 1: What should be watched? 12

    Step 2: How can it be watched? 14

        Where can the information be found? 16

    Step 3: Who should watch? 17

    Step 4: How can results be documented? 18

        Which information and for whom? When and how? . . 21

    Step 5: What was observed? 22

    Step 6: Why these results? 23

    Step 7: What action should be taken? 25

        Documenting decisions 26

        What should be changed in the monitoring system? . . 26

CASE STUDIES . . . . inside rear cover

THE TWO STRINGS OF PIM . . . . outside rear cover

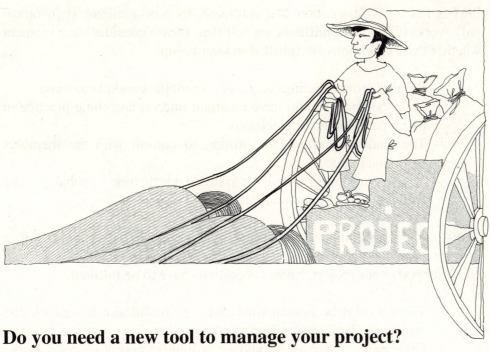

Do you need a new tool to manage your project?

Thank you for your interest in Participatory Impact Monitoring. Just call it „PIM" it's much easier!

PIM is designed for **self-help projects**. If people have set up an independent organization of their own to improve their living conditions in some way, they have a self-help project.

self-help projects

Normally, a group has leaders. Apart from them, various other members of the community may have specific responsibilities. Are you a **leader** or a **member** of a people's organization of this kind? You may call it a „self-help initiative" or a „grassroots organization". We call it a self-help group or simply a **group**.

leader

If a group is carrying out a bundle of activities to solve a specific problem in its environment, we call this a project. Are you starting a **new activity** or a „**project**"?

new activity

Your group also has **rules,** which specify how the leaders and ordinary members of the group share their responsibilities. Usually, these rules evolve from practice. Do you feel that the rules of your group need to be developed further – perhaps because a new task has been set?

rules

YES? Then perhaps you'll be interested in **Group-Based Impact Monitoring**. The purpose of PIM is to help make your self-help organization more successful by

purpose of PIM

- gearing your project activities to your members' needs,
- involving the members in observation (watching), reflection (deliberation) and decision making,
- strengthening your organization's structures.

PIM 1 · Group-based Impact Monitoring

conditions for PIM

PIM is not a magic solution that automatically works miracles! In fact it only works if certain **conditions** are fulfilled. Please consider for a moment whether these conditions are fulfilled in your group:

- Regular group meetings (e.g. once a month) should be normal.
- Your members should have a certain interest and some practice in participating in group decisions.
- The leadership should be willing to consult with the members before taking decisions.
- Your group should be willing to invest a little time – probably more than before – in joint management.

Many self-help groups are supported by a development organization, a governmental assistance organization, or an **NGO,** as we call it here. If an NGO supports your project, further conditions have to be fulfilled:

- There should be mutual trust and a mutual desire to manage the project by Participatory Impact Monitoring.
- Other actors involved should be willing to accept changes in the project, i.e. to adapt their contributions to your groups' needs.
- The facilitators should restrict themselves to methodological support; group decisions must be done by the group itself.

The NGO should then back up your Group-Based Impact Monitoring, preferably through a facilitator who should help your group to take its own decisions. The organization can also do this by applying the other string of PIM, which is known as NGO-Based Impact Monitoring. As you will see on the back cover page, PIM has two independent monitoring systems which are interlinked regularly to ensure that the NGO really is supporting what the group wants.

1. What does monitoring mean?

Let's start with an example:

example

Two farmers growing corn *(see next page)*

Farmer 1 regularly checks the plants growing in his fields. When he notices that some plants are diseased and are becoming stunted, he immediately treats them with a (biological!) remedy. His harvest is good and he is satisfied.

Farmer 2 does not look at his field while the crop is growing. At harvest time he is shocked when he realizes that most of the crop is lost. He is disappointed.

Two farmers growing corn

Monitoring means continuous **observation, reflection and correction of activities**. This is done by Farmer 1 but not by Farmer 2.

How difficult is monitoring?

ordinary activity

Monitoring is really a very ordinary activity. After all, everybody watches and thinks about things or events that are important to them: children growing, the weather, the prices of agricultural produce, the behaviour of extension officers. The people concerned usually observe better than outsiders. But sometimes an outsider's point of view may be useful.

Monitoring is easy if people themselves observe what concerns them. Monitoring can be learned: we may have to sharpen our perception, we may have to improve our choice of examples (= indicators) or the way we measure observed changes to get a better picture of what is happening.

How much information do we need?

One common mistake is trying to collect too much information.

monitoring rules

In an organization or a group, however, the activities are more complex. Observation is not more difficult, but often nobody feels responsible for monitoring. So certain rules have to be introduced for the „monitoring game". This booklet on Group-Based Impact Monitoring suggests some monitoring rules for your group.

In many projects it is helpful to introduce a monitoring system, especially if there are many tasks to be completed or different actors who need information about what is going on in a big organization. It is also useful if previous observation and reflection methods have failed or have been unsatisfactory, and a more systematic or formalized approach seems necessary.

monitoring takes time

Don't forget: **monitoring takes time!** Out of every 10 hours of work you will have to set aside about one hour for reflection (i.e. for observing and analyzing the situation, discussions with other members of the group,

making up your mind, taking group decisions, planning, budgeting, coordinating, taking notes, keeping records); otherwise you will drown in work and there will be no time left for thinking! Is it impossible?
Bear in mind that

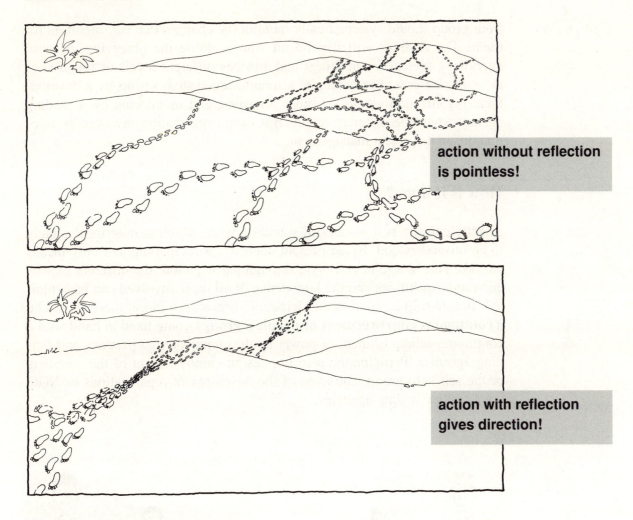

What should be monitored?

Some people who know development projects consider it important to monitor how far the plans are fulfilled. That is generally correct, because it is important for everybody to stick to their duties as agreed.

plans

But in many cases the written plans do not reflect what people really want. There can be many reasons for this:
- The self-help group and the social context are not taken into account early enough in the planning, or not to a sufficient extent.
- Only after the project and the activities get under way do people get a clear idea of what they want and how they can articulate this.
- The information basis for planning is not good enough and the working plans are incomplete.
- The general context and the internal situation of the organizations involved are changing rapidly.
- Mistakes are recognized and corrected too late.

PIM 1 · Group-based Impact Monitoring

informal objectives — This is why PIM asks about the informal objectives of the actors involved („expectations and fears") and the effects that were not planned („What has changed?"). This is detailed in the next chapter, where we describe the steps in monitoring.

changes — Your group should systematically monitor the changes that are important for them! The findings will depend on who is doing the observing, because everybody sees different things and attaches different levels of importance to what s/he sees. For example, a monitoring system set up by a development organization would set other priorities; and monitoring by a funding agency would be different again. But each organization involved is autonomous in its project management.

How is it done?

participation — „Participation" is a wonderful-sounding word which is notoriously likely to be misunderstood. In the present context, „participation" not only means „to take part in a joint activity"; through participation, the different experience and capabilities and the knowledge of **all** those involved can be exploited. In self-help promotion, participation means even more: there should be

continuous empowerment — a **continuous empowerment of people's groups** going hand in hand with a continuous relinquishment of power by development organizations and funding agencies. Participation also implies an empowerment of the members of the self-help group, and even of the development organizations or NGO vis-à-vis the funding agencies.

Participation means continous empowerment of partners

> **Participation is an ongoing process which requires ongoing changes.**

Participation is thus an ongoing process where one side discovers capacities of its own and learns to act more and more autonomously, and the other side learns to accept other viewpoints and to hand over responsibilities and power.

Learning processes and capacity-building are not feasible without action. We are „learning by doing", and this again means learning by **trial and error**. It may be helpful for all actors involved to know that:

> **Mistakes are useful and often necessary for learning.**

Participatory monitoring should help those involved to learn to draw conclusions for decision-making out of this trial-and-error process and guide the activities according to „lessons learned".

In PIM, participation has one more meaning. We stated above that the various actors each have their „own" project managements. The most important actor is the self-help group and their experiences must be preserved. In the case of a joint project these **different experiences and monitoring systems must be linked**. There should be an exchange of information and joint reflection – this is the basis of participatory monitoring.

So is PIM always participatory? No management tool can be participatory in itself – not even participatory monitoring is participatory *per se*. **The utilization of a tool is an art, and requires certain special attitudes:**

> **Participation requires mutual acceptance, openness, trust and confidence.**

2. Steps in introducing and carrying out Group-based Impact Monitoring

As we have seen above,
PIM holds out the promise of being useful – but certain conditions must be fulfilled.
PIM is an appropriate tool for managing a self-help project – but do you really want to try a new monitoring concept?
PIM may take more time – but do your group members want to spend more time on joint decisions?
PIM is intended to empower people who have no voice – but will the people in power at the moment accept a loss of influence?

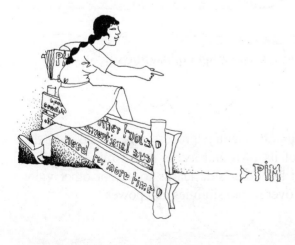

Is your organization strong enough to tackle this new task? Are the leaders and members willing to face the hurdles mentioned above? If so, it is very likely that you will be rewarded with fruitful growth of your activities!

PIM 1 · Group-based Impact Monitoring

Steps in introducing PIM — STEP BY STEP

1. What should be watched?   Expectations and fears of the group members

2. How can it be watched?   Concrete examples (indicators)

3. Who should watch?   Elected group members or an already existing committee of the group

4. How can results be documented?   Graphs, charts, descriptions, that the group can understand, as well as other forms compatible with verbal transmission culture

Steps in carrying out PIM

5. What was observed?   Reports at the beginning of group meetings

6. Why these results?   Assessment and analysis by the group

7. What action should be taken?   Immediate decision making at the meeting (= plan adjustment)

Preliminary step: What is known about the context?

If your group prepares a joint project with other groups or organizations, you should be aware of the context:

- **What does the group know about the problems you want to resolve?**
  (= situation analysis)

situation

Before starting new group activities, the members of your group should have time and an opportunity to think about their own situation: the problems you want to tackle, as well as your strengths, your weaknesses, your opportunities and the hazards.

- **What does the group know about the other actors associated with the project?**
  (= participant analysis)

other actors

You should also look at other organizations and people associated with your project. Do they have similar problems, similar interests, similar ideas ...? You may want to make a list: who would be affected by the project, and who could take over some of the tasks? What are the various individuals' main responsibilities? What are their strengths and weaknesses? What role could each actor play?

- **What do the partner organizations involved know about you?**

exchange of experiences

External organizations use to have their own monitoring systems. However, your group should regard group-based impact monitoring as your own independent monitoring system.

If your group already has a good management and decision-making structure, whether formal or informal, you should explain this structure to the external organizations involved. Otherwise they may think your organization has no rules. If your rules work well they will be respected and there is no need to change them; on the contrary, the external organizations should profit from your experience! On the other hand you may want to improve them in which case dialogue with your partner organization could be fruitful.

PIM 1 · Group-based Impact Monitoring

Step 1: What should be watched?

expectations and fears

When new project activities start, one of the regular group meetings should be set aside to identify the expectations and fears of the group members. At the meeting a simple collection of ideas can be made on what people expect from the project and what reservations they have about it.

- **What changes do we expect from the project?**
  (What do we hope to achieve?)
- **What changes do we fear from the project?**
  (What do we want to avoid?)

There will be various expectations and fears. In the case of the shop that was to be run by housewives' committees in Caracoles, Bolivia, the following concrete examples were given:

example

EXPECTATIONS	FEARS/DOUBTS
that the nutrition of the cooperative's members would be improved	*the consumption of highly nutritive food products has increased*
that basic foods would be available at fair prices	*that cooperative members would not pay their debts*
that housewives would participate actively	*that they would not be able to run the shop and would go bankrupt*
that the housewives would learn how to run a shop	*that they knew nothing at all about administrative procedures*

Out of these expectations and fears the group selects those which are most relevant, for example by prioritizing. About three expectations/fears may be sufficient. The expectations and fears thus chosen form the object of observation in group-based impact monitoring.

The list of things to be watched should not be too long. Generally, a few things (perhaps even just one?) which can be analyzed cooperatively are clearer than a long list of complex facts which nobody can memorize or process. The following picture illustrates how complicated and time-consuming an excess of information can be.

How much information is needed to find the way?

Little, but crucial information: *Too much information:*

Step 2: How is it done?

After having chosen some expectations or fears, you should ask for concrete examples of how you can see if things change the way you want or not. You are looking for indicators.

EXPECTATIONS/FEARS	INDICATORS
that the nutrition of the cooperative's members would be improved	the consumption of highly nutritive food products has increased
that basic foods would be available at fair prices	prices have been linked to market prices
that housewives would participate actively	housewives' committees have been informed about the administration of the shop at each meeting
that it would not be supervised by the housewives' committee	
that cooperative members would not pay their debts	a register is being kept of names of cooperative members who have not paid their debts

Indicators are like markers

Indicators are rather like roadside markers: they show whether you are still on course and what progress you have made.

indicators

Generally, it should not be too difficult for your group to provide some good examples and thus to define the indicators during the meeting itself. The group members not only know their world in detail, they often have their own indicators for assessing changes relevant to them.

Concrete examples of the expectations and fears may also be identified with the help of a facilitator, if you need assistance.

PIM does not require scientific solutions; it asks for practical solutions for self-help groups and small development organizations/NGO. Continuous reflection is more important than meticulous gathering of data.

practical solutions

Very briefly, we would suggest four ways to establish indicators. Choose the easier ones!

PIM 1 · Group-based Impact Monitoring

FOUR WAYS TO CREATE INDICATORS

1. Measuring or counting: gives us exact numbers
   example: the prices of the goods are measured: $ 0.45 per kg;
   the quantity is counted: 20 pairs of shoes

2. Scaling or rating: gives us a gradual description
   example: the quality of the goods can be scaled:
   very good --- good --- average --- bad --- very bad

3. Classifying: informs about non-gradual categories
   example: Is salt available: yes / no ?
   Who takes the final decisions in the shop: women / men ?

4. Describing qualitatively: describes in words only
   example: How is the shop administered? The answer describes one or more aspects in words. The description may have a certain structure: (positive - negative aspects; leadership - purchases - sales - book-keeping)

Where can the information be found?

own sources, knowledge of group members

We should determine beforehand where and how we find the information for the indicators we have chosen. Collecting data takes time and costs money. Information from our own sources is more authentic and may initiate more discussions. In PIM, we basically trust in the knowledge of group members themselves. It takes less time to obtain outside information but it may not be so relevant. If outside information is used the source should be quoted.

example

INDICATORS (derived from expectations or fears)	OBSERVATION METHODS
the consumption of highly nutritive food products has increased: - higher consumption of lentils and quinua - lower consumption of noodles, rice and sugar	measure the weight of each product sold (lentils, quinua, cornflour, dried field beans, noodles, rice and sugar) separately, and record it appropriately
prices have been linked to market prices by agreement with the three cooperatives	reply YES or NO and record observations and comments
women have been informed monthly about the administration of the shop at each Housewives' Committee meeting	reply YES or NO and record observations and comments
a record is being kept of the names of cooperative members who have not paid their debts	keep the record in the form of a table, with a separate table for each cooperative

Step 3: Who should watch?

The people responsible for watching these indicators should be chosen at the meeting. The group could nominate individual observers or an „observation team" or „committee". These people should have a specific function within the organization, to avoid duplicating structures.

By assuming the role of observers, members of the group learn to watch for relevant changes and to assume responsibility.

observation team

Even two or more observation teams might be useful if different social groups perceive the project differently or benefit from it in different ways (e.g. women/men, the younger/older generations).

If the task at first seems too difficult for the group, the NGO's field staff may be asked for assistance. Or someone from the community might assist, for example elders, informal leaders, former leaders, teachers, etc.

Step 4: How can results be documented?

In day-to-day life you do not normally document what you have observed. So why might it be important to keep a record?

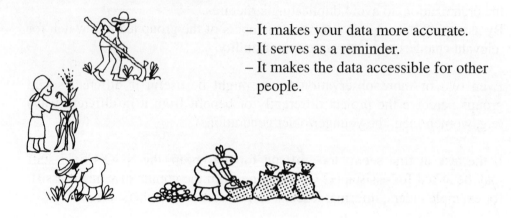

- It makes your data more accurate.
- It serves as a reminder.
- It makes the data accessible for other people.

example

The following graph demonstrates how keeping records can be helpful and useful:
- Three observers (on the next picture on the left) watch fruit being sold at a village market, but without taking notes; at the end of the day they cannot agree on how much was really bought and sold; consequently they cannot take a decision regarding an appropriate means of transport.
- Two observers note down immediately what they have seen; at the end of the day their results are very similar, so it is easy for them to decide which method of transporting their fruit is the most appropriate.

WHY DO WE HAVE TO TAKE NOTES?

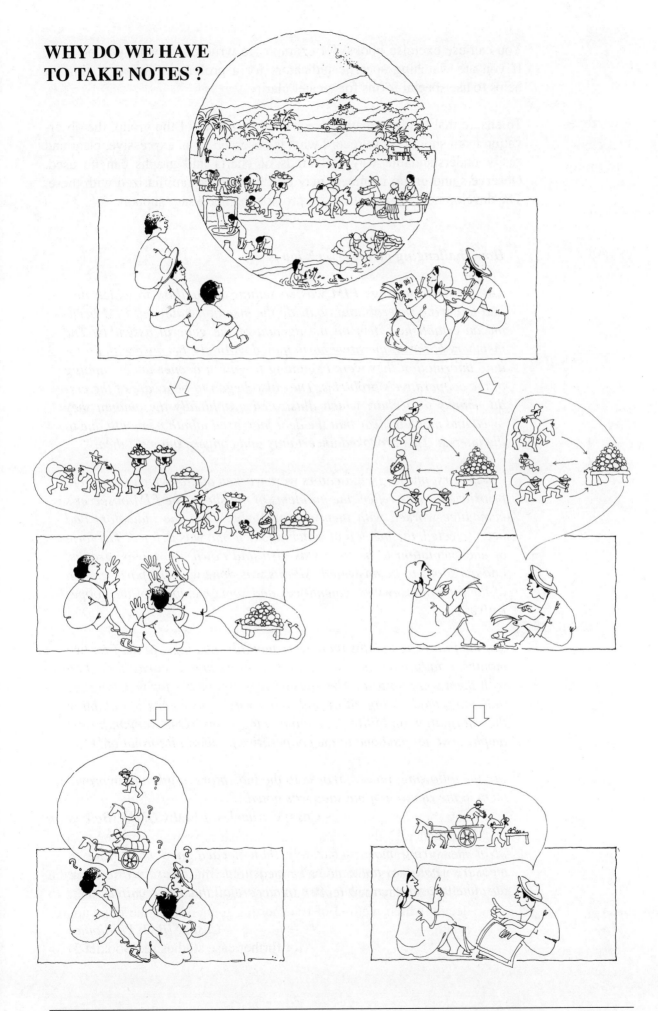

PIM 1 · Group-based Impact Monitoring

You can use exercise books, for example, to write down what you observe. If you are watching specific indicators for a prolonged period of time, it helps to use special forms for greater clarity.

recording in a clear and understandable way

To ensure that the information reaches all members of the group, the observation team should try to record what is observed in an expressive, clear and easily understandable way. All kinds of charts and graphs can be used. Observers and group members may first have to be familiarized with these. The development organization can provide the necessary support.

example

How challenging an observer's job is

„At the first meetings PIM was something unfamiliar, to which they had to devote several hours outside the meeting. Later on, PIM became an established item on the agenda of the general assembly. The members of the cooperative participated willingly, because with up-to-date information they were beginning to gain a deeper understanding of the cooperative's problems. They also began to be aware of the credit, thanks to a chart which illustrated graphically the amount they owed and made it clear that the debt increased month by month due to the interest. This provoked uncertainty and concern amongst them."

„In the first phase, 14 indicators were worked out, of which eight were prioritized: these reflect the problems of the shop. The PIM observer's committee worked with them continuously. When the indicators had been selected, the method of watching each one was defined. The form of documentation to be used was also laid down, i.e. tables, graphs, and questions to be answered. All this was done with the participation of the three housewives' committees and some members of the cooperative."

„In the following months there were monthly meetings. In the first two months, a failure to understand the indicators and mistakes in dealing with them were noticed. The housewives responsible for watching the indicators made a big effort and found ways and means of obtaining the information and filling in the indicator forms. (One woman, for example, sent her husband to the cooperative to obtain information!)"

„In the following months, thanks to the indicators, a gradual improvement in the running of the shop was noted."

Case: Caracoles, FEDECOMIN, Bolivia

„Not measuring more accurately than needed, and not trying to measure what does not need to be measured. We are trained to make absolute measurements, but often trends are all that are required."

Case: SIBAT, Phillipines
(for further case studies see Booklet 3)

Which information and for whom? When and how?

What should be done with the information? Not everyone needs all of it. Information from inside the group does not necessarily have to be given to outsiders, to development organizations (NGO etc.), funding agencies, or other external actors (government authorities, intermediaries etc.). You should agree within your group, and with your (external) partners as well, what information is to be kept inside or channelled outside, and how often.

The group probably needs frequent reports. These can be presented at the regular meetings.
Not all information is required monthly; some is needed less often.

If your group has agreed on an information flow with external partners, this exchange need not be frequent. The further away the other actors (NGO, funding agency) are, the lower the intensity of reporting will be. In the long run, they may require information only at quarterly or annual „joint reflection workshops" (see NGO-Based Impact Monitoring).

So far, we have focused on setting up the group's monitoring system. Now we shall see how regular group-based impact monitoring can work.

Step 5: What was observed?

relevant changes

The members of the monitoring team have now started to keep a watch and document the indicators. At the beginning of a group meeting they should report their findings, i.e. the relevant changes within and around the group. This can be backed up by informing the entire group beforehand, with posters or other public information.

If there are a large number of indicators, the purpose of the meeting should be taken into account and the indicators chosen accordingly.

Following this presentation the members of the group may be asked if any other unintended and unexpected changes have occurred. Asking questions may result in other relevant changes being identified which are unrelated to the expectations and fears mentioned above. (If there is an external facilitator this step will be easier.)

Step 6: Why these results?

The results should be briefly discussed and analyzed. If the indicators only require YES or NO as an answer, it is useful to ask for further findings („Why is it yes or no?") and to give details.

discussion and analysis

Indicator 2:	*Were the prices linked to market prices, in agreement with the three cooperatives? (yes/no)*
Answer:	*YES*
Observations:	*Prices were adjusted according to the increase in market prices, including an amount for transport and travel expenses. The housewives from cooperative N. did not attend this meeting.*

example

It is probably not advisable to start a discussion on each finding immediately, because on many details there will be no consensus. A discussion should only be started if there is a need to take decisions, to act or respond to a change or to persistent problems. If the results are felt to be „normal" there is no need to discuss them. However, if they deviate from what was expected it may be necessary to give the matter some thought and take a decision.

„Traffic lights" to indicate the need for decisions

Example: The supply of goods is expected to be regular; observations could be:

goods are normally available:	**green:**	*no additional action necessary*
an item was out of stock for 7 days:	**amber:**	*corrective action by stock-keeper necessary*
an item was out of stock for 30 days:	**red:**	*special meeting and decision required*

PIM 1 · Group-based Impact Monitoring

The indicators make the findings fairly easy for other members of the group to check, but they will not always be accepted without any doubt. If there is any disagreement it may be important to point out that change in the community can be (and usually is) perceived in different ways.

Generally, the results of observation require analysis above all in the following cases:

- If things are always as expected, this is probably a success and it is worthwhile analyzing occasionally why and how these results have been achieved.
- If the monitoring results show that there are problems which require decisions, the meeting should put the topic on its agenda immediately.

causes and consequences of changes

The findings can then be analyzed on the basis of the monitoring reports, either at the same meeting or at one of the next meetings. It is important to discuss the consequences and causes of the changes noted and their relevance to the future activities of individuals and of the group.
It is important to ask:

- **What has been (can be) influenced by <u>ourselves</u> (the group)?**
- **What has been (can be) influenced by the <u>development organization</u>?**
- **What has been (can be) influenced by <u>others</u>?**

If the changes are regarded as good, this will give you the self-confidence to demonstrate that your personal and collective activities have been significant and successful. If the changes are regarded as bad, your group should look for possible ways of influencing them positively, either by taking action itself or by urging others to do so.

PIM is action-oriented!

There is a certain risk of becoming embroiled in pointless discussions at this stage of data assessment and analysis. To avoid wasting time and effort, a facilitator should try to structure the debate and to keep it action-oriented.

Step 7: What action should be taken?

The impact analysis has raised the question „what can we influence?". The group should take decisions accordingly. If possible, you should decide immediately, at the meeting, what initial action should be taken. This shows how PIM makes sense. Using PIM in practice, it has been found that decisions are best taken immediately after discussing the indicators.

„Since all of the activities are geared to enabling members of the group to bring about changes in the community, provision must be made for allowing them to state where, in their view, fresh action is needed and what should be done to complete it.
There could even be a situation where the measures taken by each member would suffice and no further action would be needed."
                                                            Case: SIBAT, Philippines

example

A plan of action should at least contain the following questions:

– What action needs to be taken?	*Cooperative treasurer should be persuaded to deduct debts from payments to cooperative members and pay them directly to the shop.*
– Who is responsible for taking it?	*Anna.*
– Who else is involved?	*Berta and Clara talk with Cooperative's President.*
– When should it be done by?	*30 April.*

example

PIM 1 · Group-based Impact Monitoring

Documenting decisions

findings and lessons learned

As a general rule, all decisions taken at group meetings are recorded in the minutes. The findings of the monitoring committees and/or observation teams and the changes as perceived by the rest of the group should also be documented. Comparison of indicators over a longer time-span will be a valuable aid to evaluation in the future. It is important to ensure that the lessons learned in the project are not lost.

sharing of information

These results are usually documented in the group's books or files. However, the sharing of information among all members of the group is part and parcel of PIM. Documentation should therefore be publicly accessible, ideally in the form of posters and small publications.

What should be changed in the monitoring system?

After a certain time you may need to adjust your monitoring system. There can be many reasons for this:

- If you notice that the indicators you chose first are not very helpful, you should look for other (or additional) indicators, in the light of your experience so far.

- In the course of time new expectations and fears may arise. More attention may be given to previously neglected changes in the group's environment. A routine will become established, of observing changes, analyzing cause-and-effect relationships, and taking decisions based on monitoring results. This will lead to improvements in the impact monitoring system.

- Development organizations and funding agencies will feel a certain need for improved information flow. They will appreciate autonomous group-based monitoring and they will support it. However, at the same time they may ask for more information from the group. The group must decide, but also negotiate with the external organizations, how (and which portions of) its information should flow outside, and through which channels.

In the long run, a self-confident group may take the interests of the development organization and funding agency into account and include them in group-based impact monitoring. If compromises are possible, group-based impact monitoring may even replace most external monitoring.

CASE STUDIES

To help readers understand the context of the quotations in the text, the five case studies on which this PIM booklet is based are briefly outlined below.

Case: SIBAT, Philippines
Community of Bacgong, Philippines

The Agricultural Production Project was initiated by PARTNERS (an NGO member of the SIBAT network). Its aim is to support and sustain socio-economic development initiatives in villages, one of which is Bacgong. The self-help group of this village comprises three sectors: farmers, women farmers and youth.
The project was conceived by these groups and PARTNERS to meet the villages' most basic need for food. The project has two major components: 1) organization and education, and 2) agricultural production, comprising rice, vegetable and livestock production mainly for home consumption.

Case: Kantuta, FEDECOMIN, Bolivia
Ore Dressing Investment Project, Kantuta Mining Cooperative, Bolivia

FEDECOMIN is the Mining Cooperative Federation of La Paz, with 76 member cooperatives. FEDECOMIN has its own advisory service, i.e. CODECOMIN.
The mining cooperatives are a genuine form of self-help, but the miners' families are very poor. The Kantuta Mining Cooperative has invested thousands of dollars in its ore-dressing operation, but it has management problems and is being hampered by delayed approval of the credit.

Case: Caracoles, FEDECOMIN, Bolivia
Housewives' Store, Cooperative Sector of Caracoles, Bolivia
(Mining Cooperatives of Libertad, Porvenir, El Nevado)

The Cooperative Sector of Caracoles consists of three mining cooperatives: Libertad, Porvenir and El Nevado. In each one the women (who are generally not members of the cooperative) have formed housewives' committees. With the support of FEDECOMIN, the three housewives' committees have established a „People's Consumption Store" in order to improve the supply of everyday goods to these remote mines and to strengthen the women's position. But the store does not work very well.

Case: PWDS, India
Community-based Candy Making, Tamil Nadu, India

Palmyrah Workers Development Society (PWDS) promotes the socio-economic development of the palmyrah tappers and other weak groups in the community.
Normally, tappers sell the raw product to merchants or the women process it in their homes. This involves a lot of work, but generates little income. In the Candy
Production Units, the tapper families have started joint production of high-quality candy.

Case: INDES, Argentina
Rural Women's Group „Unión y Progreso", Misiones, Argentina

INDES is an NGO working in the province of Misiones with about 20 farmers' groups, two of which are women's groups. Within 5 years, Unión y Progreso's membership has increased from 25 to 83. Its objective is to improve the families' living conditions. The main projects are vegetable gardens for family consumption, poultry production, training courses (health, nutrition) and a credit programme; the women hope to produce surpluses which they will be able to sell.

Dorsi Germann / Eberhard Gohl

Participatory Impact Monitoring

Booklet 2:
NGO-Based Impact Monitoring

A Publication of Deutsches Zentrum für Entwicklungstechnologien – GATE
A Division of the Deutsche Gesellschaft für Technische Zusammenarbeit (GTZ) GmbH

The authors:

Dorsi Germann, sociologist and graphic artist, spent four years working in a community development project in Senegal. For the last fourteen years, she has been a consultant on adult education, appropriate technologies, technics of communication and visualization, project management, monitoring and evaluation, organizational development and participatory methods in Africa, Asia and Latin America, mainly working for GTZ and FAKT.

Eberhard Gohl, economist and sociologist, spent a few years in Turkey, Peru and Bolivia. For eight years, he worked mainly with FAKT, DSE-ZEL and GTZ as a consultant for project management and organisation development. At present, he works as Controller in the German Protestant Churches' funding NGO „Bread for the World".

Die Deutsche Bibliothek – CIP-Einheitsaufnahme

Participatory impact monitoring : a publication of Deutsches
Zentrum für Entwicklungstechnologien – GATE, a division of
the Deutsche Gesellschaft für Technische Zusammenarbeit
(GTZ) GmbH. – Braunschweig ; Wiesbaden : Vieweg.
  ISBN 3-528-02086-5
NE: Deutsches Zentrum für Entwicklungstechnologien <Eschborn>

Booklet 2. NGO based impact monitoring / Dorsi
  Germann/Eberhard Gohl. – 1996
NE: Germann, Dorsi

The author's opinion does not necessarily represent the view of the publisher.

All rights reserved
© Deutsche Gesellschaft für Technische Zusammenarbeit (GTZ) GmbH, Eschborn 1996

Published by Friedr. Vieweg & Sohn Verlagsgesellschaft mbH, Braunschweig/Wiesbaden

Vieweg is a subsidiary company of the Bertelsmann Professional Information.

Printed in Germany by Lengericher Handelsdruckerei, Lengerich

ISBN 3-528-02086-5

CONTENTS

Does your NGO or development organization need
a new tool to manage its projects? 3

1. General Ideas on Monitoring 4
1.1 Three types of organization 4
    Participatory management 7
1.2 Monitoring 8
    Purposes of monitoring and evaluation 9
    First approach to monitoring: emphasizing periodical reflection ... 10
    Second approach to monitoring: based on evaluation 10
    How can monitoring be systematized? 12
1.3 Impact Monitoring 13
1.4 Participatory Monitoring 17
    Participation 17
    Participation and monitoring 18

2. NGO-Based Impact Monitoring 19
2.1 Advantages and obstacles 19
    Introducing PIM to NGO field staff 21
2.2 Steps in introducing and carrying out NGO-based
monitoring of socio-cultural impacts 22
    Preliminary Step: What do we know about the context? 22
    Step by Step
    Step 1: What should be watched? 23
    Step 2: How can it be watched? 25
    Step 3: Who should watch? 27
    Step 4: How can the results be documented? 28
        Which information and for whom? When and how? ... 28
    Steps 5 to 7: What did we observe?
        Why? What should be done? 29
2.3 Joint Reflection Workshops 30
    Formulation of guiding questions 32
    Workshop Step 1: What has changed? 33
    Workshop Step 2: What have people learned? 34
    Workshop Step 3: What action must be taken? 34
    Workshop Step 4: How can we improve our
        impact monitoring? 35
    Post-Workshop Step (5): What conclusions can we draw
        for our work? 35
    Concluding remarks 36
2.4 Facilitating the PIM process 36
    How can PIM be introduced into the
people's organization or self-help group? 36

Does your NGO or Development Organization need a new tool to manage its projects?

Thank you for your interest in Participatory Impact Monitoring. Just call it „PIM" – it's much easier!

PIM is designed to make self-help projects and organizations more successful by

- gearing project activities to the self-help group members' needs

- involving members in observation, reflection and decision-making

- strengthening the organization structures.

Booklet 1 is written for leaders or members of self-help groups and describes how **group-based impact monitoring** works.
Booklet 2, on **NGO-based impact monitoring** is addressed to staff members of development organizations, i.e. national organizations such as NGOs, federations or government organizations which promote self-help groups.

- Do you need a monitoring instrument to manage your projects?
- Do you want to document the socio-cultural impacts of your work?
- Do you have to justify the success of your work perhaps because of some technical or economic error?

- Do you want to improve interaction between the self-help groups and your own?
- Do you need more information on learning processes within the self-help group and your own?
- Do you want more transparency in your organization and decision-making structures?

If so, why don't you read this booklet and try to implement PIM!

conditions for PIM

PIM is not a magic solution which works automatically. In fact, it only works if certain **conditions** are fulfilled. Please consider for a moment whether your NGO fulfils these conditions:

- Your team should be willing to promote people's participation, i.e. increasing the autonomy of the self-help group.
- There should be mutual trust and a desire to manage the project transparently by participatory impact monitoring.
- Your team and other organizations and individuals involved should be willing and able to accept changes in the project, i.e. to adapt your plans and contributions to people's needs.
- Your team should also be willing and able to invest a little time – probably more than before – in monitoring. (You will then avoid wasting time on pointless activities.)

The self-help group should also fulfil certain conditions; they are mentioned in Booklet 1.

In PIM, two independent monitoring systems are regularly interlinked to ensure that the NGO really is supporting what the self-help group needs and wants.

Your NGO-based impact monitoring, which is outlined in this booklet (no. 2), should therefore be linked to the autonomous group-based impact monitoring described in Booklet 1.

1. General Ideas on Monitoring

1.1 Three types of organization

In self-help promotion, we distinguish three main types of organization:

- 1. **self- help groups** or people's organizations   (GROUP)
- 2. Development organizations,
    NGO or self-help support organizations   (NGO)
- 3. **Funding Agencies** or donors   (FA)

Each of them sees a certain set of problems, which can be solved by a combination of activities. This combination of activities is called a project. Each organization (i.e. each actor) has „its" project, so there are at least three different projects. The basic interface linking them is the project agreement.

Marketing Project example

Project agreement: Storehouses and marketing assistance for cereals and potatoes; training courses on food processing and nutrition.

Each of these organizations is interested in monitoring and managing „its" project. Naturally, each has a different point of view:

> The **groups** are primarily interested in tangible improvements in their living conditions. They want a storehouse in the village, or better prices (technical and economic impact). They may want to be able to negotiate more effectively with the intermediary, or to know more about storage and processing of their products (socio-cultural impact). However, aspects concerned with „capacity building" are not normally mentioned explicitly, although it may be felt necessary to link them to technical and economic changes.

> **NGO** and **funding agencies**, on the other hand, are interested in the long-term effects of their assistance. Of course they want to offer immediate assistance to the people to improve their living conditions, but for them self-help projects have only one overall goal, i.e. to enhance people's capability to act and in the long run to help themselves.

While these views are not contradictory, they are not congruent either. To some extent the projects of the various organizations overlap.

Participatory management

Three types of organizations are involved in management. But how deeply is each participant involved? What does participation mean in this context?

Often, in reality, the cooperating partners are very different. They own different amounts of money or power or have different educational, social and cultural backgrounds. There are multiple communication problems (see picture).

need for good communication

said is not yet heard

heard is not yet understood

understood is not yet approved

approved is not yet applied

applied is not yet continuously applied

continuously applied is not yet being satisfied

To find joint (and most appropriate) solutions to problems which have been identified, we need to listen to each other and to learn from each other, and we need mutual acceptance, mutual trust and confidence. We need motivation, commitment, creativity, and good and open cooperation between all people concerned.

participatory attitudes are a precondition

Participation in this context means that each group of actors is autonomous in its decision-making, and the actors have to reach agreement on a basis for cooperation. The result must be joint action, and the principal actors should be the people themselves. If the outsiders take the lead, there is an enormous risk that the self-help group will be nipped in the bud. This is why **participatory attitudes** are a precondition for the work of DOs and FAs. They need knowledge and experience in using participatory techniques.

> Participation does not come naturally. No management tool can be participatory in itself. The utilization of a tool is an art! Participatory methods have to be developed and learned, and PIM may also be regarded as a tool for learning.

implications with regard to the actors:

- We assume that the actors belong to one of three standard categories:
  - self-help groups or people's organizations,
  - NGO or development organizations,
  - funding agencies.
- Each group of actors has „its" project which it wants to manage.
- These projects probably overlap but are not congruent.
- In self-help projects, the principal actors should be the people themselves.
- „Participatory Management" can be learnt.

1.2 Monitoring

monitoring: observing, collecting information, reflecting.

Monitoring means observing and collecting information, and reflecting on what has been observed, to check whether we are still „on course" to achieving our aims and if necessary to change course. It is like navigating a ship (our project) between reefs and through shallow water towards an attainable goal.

As we have seen, all the organizations involved carry out „their" projects. Consequently, in one way or another, they also perform their own monitoring. But their monitoring is often felt to be unsatisfactory: perhaps it is not systematic, or too time-consuming, or data are „not available".

Purposes of monitoring and evaluation

How can monitoring and evaluation contribute to management? Generally, monitoring has two purposes:

1. **checking:** Does everybody carry out her/his duties as laid down in the project agreement or in the adjusted plans? What is the ratio of input to output (efficiency)? How good is the outcome (effectiveness)?

2. **reflection and learning:** What can we learn from our successes and failures? What have we learned to do ourselves since we started? To what extent are we capable of helping ourselves?

emphasis on reflection and learning

In the participatory approaches to self-help promotion, the emphasis is naturally on reflection and learning. Checking is primarily understood to be self-oriented.

Briefly, by **two approaches**, we will try to explain how monitoring is understood:

First approach to monitoring: emphasizing periodical reflection

We are all familiar with the **action – reflection – action** sequence:

self–critical assessments

In simple terms, it involves a continuous flow of activity and a self-critical assessment of our actions. This implies that a self-critical assessment should be made before and after each important action.

In an organization or a group there are more activities, which may also be more complex: the periods of reflection will vary according to the organization and the decision-making level.

example

*Each team member probably reflects **daily** on the activities he performs within the organization. The team may reflect **weekly** on its duties. Other meetings at departmental level may be held **monthly**. And **once a year**, there may be a general review of the long-term concepts of the whole organization (and **every 6 years** an external evaluation by a funding agency?). Each organization makes its own rules for these reviews: of course, they may also be irregular or have quite different time-spans.*

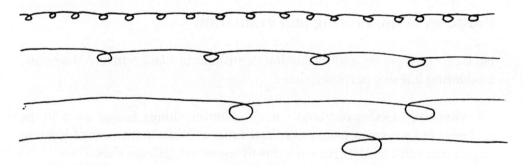

Second approach to monitoring: based on evaluation

Many of us are familiar with **monitoring and evaluation**. Clearly, they differ in depth and periodicity (see above). But apart from that, what exactly is the difference? Let's try the following definitions:

monitoring: a process of systematic and critical review of an operation with the **aim** of **checking operation and adapting it** to circumstances.

This implies that monitoring is a more frequent form of reflection, mainly at operational level, subject to a limited range of decision-making.

evaluation: involves comprehensive analysis of the operation with the **aim** of **adapting strategy and planning** to circumstances.

This implies that evaluation is a less frequent form of reflection, it is deeper and leads to more fundamental decisions.

In our example „Two farmers growing corn" (see booklet 1) we compared two farmers:

Farmer 1 observes the plants growing in his fields at regular intervals. When he notices that some plants are diseased and are becoming stunted, he immediately sprays them with a (biological!) remedy. His harvest is good and he is satisfied. Farmer 2 does not look at his field while the crop is growing. At harvest time he is shocked when he realizes that most of the crop is lost. He is disappointed.

In this example, monitoring as continuous observation and correction was practiced by Farmer 1, not by Farmer 2. After the harvest, both made an evaluation and reflected on their strategy and plans for the next period. But Farmer 2 made the evaluation only after a failure and without having monitored.

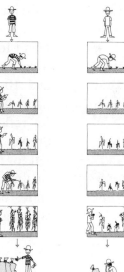

As monitoring and evaluation are thus two sides of the same coin (often referred to as „M+E"), differing only in frequency and range of decisions, monitoring often goes hand in hand with evaluation.

M+E
two sides of the same coin

However, we believe that it is useful to separate them for the purpose of analysis, and for greater flexibility and more conscious evaluation. The following activities are **typical elements of monitoring:**

ongoing review	to observe changes in project implementation,
systematic documentation	to document this process of change,
analysis and decision-making	to reflect, to adjust and to rectify the operation.

PIM 2 · NGO-based Impact Monitoring

> **Monitoring: frequent and systematic periodical reflection**
>
> Monitoring means frequent and systematic periodical reflection in order to manage a combination of activities of an organization or a group. It is self-conducted, and promotes continuous development of personal capacity, team development and organizational development.

How can monitoring be systematized?

no need for too many data

If you try to observe too many things, monitoring can take up too much of your time. A few things may be sufficient, and can be documented and analyzed jointly and more easily (see example in booklet 1). Often you only need to know the trends.

personal reflection is valid

Each organization formulates its own rules for periodical reflection. All individual forms of personal reflection are valid.

reflection takes time

Don't forget: **reflection takes time!** You have to set aside between 5 and 10 per cent of your working hours and a portion of your budget for reflection – otherwise you will drown in work and have no time left for thinking!

reflection saves time

In this way, **reflection also saves time**, because it helps you avoid wasting time on pointless activities.

> **Implications for the monitoring concept**
>
> - monitoring in self-help promotion encourages reflection and learning
> - periodical reflection is necessary before and after action
> - a clear distinction between monitoring and evaluation is not necessary
> - we have chosen the term monitoring to emphasize frequent reflection with
>   - permanent observation
>   - systematic documentation
>   - and finally, the taking of decision continuously during the action
> - you have to allow some time of reflection – to avoid wasting time on pointless activities!

Many projects will benefit from the introduction of a monitoring system, especially if the tasks are complex or if, in a complex setup, there are various actors who need information on the processes. It is also useful if the previous reflection or monitoring concepts have failed or have been unsatisfactory and a more systematic or formalized approach seems necessary.

action without reflection is pointless!

1.3 Impact Monitoring

What do you want to monitor?

You can monitor

1. the **budget:** this is very important but requires specific instruments. To describe them here would be beyond the scope of this publication.
2. the **activities:** sometimes, if people are not used to responsible and independent work, you have to check what they are doing all day; but that is not the task of participatory impact monitoring.
3. the **project objectives:** then we refer to planning documents, and may have indicators for **results, project purpose** and **overall goals**, as well as for key **assumptions or risks**.

The last-mentioned example (3.) referred to a logical framework suitable for setting up a conventional monitoring system. With this approach, impacts are understood as lasting and significant effects at the level of the overall goal. They are only identifiable some time after completion of a project.

We don't think it is helpful to use the term „impact" in a restrictive sense. To define an impact monitoring concept it is better to include a wider range of interpretations of „impact".

> Pause here for a moment before you read on: What do **you** understand by „impact"?

All subjectively important changes associated with the project may be impacts. All of the individuals, groups or organizations involved have their own reasons for taking action, their intentions. These intentions represent informal objectives, which guide the activities of each actor. The more informal the context of the actor concerned, the more relevant these intentions. Subjectively important changes are far more essential than formally fixed goals.

importance of the informal side of organizations

Therefore, each actor has, consciously or unconsciously, her/his own monitoring to manage her/his activities. In self-help promotion, the people's expectations and fears concerning the project are often concealed. In cultures where informal communication is the norm it is not easy to articulate oneself on a formal communication level.

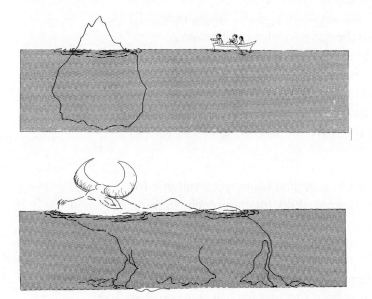

More formalized organizations, like NGO, other development organizations and funding agencies, also have their informal side. Often, many of the personnel's objectives, and possibly also those of the organization, are not clear. The famous iceberg illustrates that only 1/7 of the organization is visible, while 6/7 is invisible.

At one workshop we asked: „**What do you feel important to observe?**" Here are some answers of different actors which we received:

self-help groups	NGO / development organizations	funding organizations
tangible results	planned results	planned results
technical learning	availability of resources	correct use of funds
respect for own needs	relation to counterparts	grassroots' participation
respect for autonomy	all kinds of difficulties	self-help processes
personal respect	grassroots' participation	

high motivation by subjectively important objectives

Monitoring is often seen as an unpopular, time-consuming activity requested by the funding agency. Planning matrices, highly abstract goals and objectives do not directly equate with the felt needs and expectations of the people concerned. **Relating to the objectives which are important to them increases the motivation for action, for active management and for monitoring.**

We suggest that impact monitoring should focus on subjectively important changes. When introducing an impact monitoring system, when defining „What to observe or to 'monitor'?" each group of actors should ask: „Which changes are or will be important to us?"

focus on subjectively important changes

The answer to this question generally takes us a lot further than the formal planning documentation, because precisely these subjective aspects are relevant to people's actions. If the formal plans are good, all these expectations are already laid down in the planning documents. Normally, however, a formal plan is an agreement between various actors, a compromise which cannot fully reflect what each partner wants. In many cases, housewives, villagers, craftsmen, workers etc. are not used to speaking out at planning meetings, simply because they normally express themselves in culture-specific ways which are completely different from planning codes. This is the reason why project planning generally fails to overcome the handicap of intercultural misunderstandings. However, new participatory methods for situation analysis and planning are helpful in giving people a voice.

objectives based on planning documents

subjectively important objectives

One more distinction should be made:
- **Technical and economic impact** covers all physical, technical, economic and financial changes; the technical and economic impact is relatively easy to observe and measure.
- **Socio-cultural impact** covers learning processes and changes in behaviour and attitudes; the socio-cultural impact is relatively difficult to observe and measure.

Most conventional monitoring systems focus on technical and economic impacts (outputs, results). However, the essential change self-help projects aim for is an increase in the group's capability to act autonomously.

monitoring of socio-cultural impacts

It is impossible to distinguish impacts methodologically; they are interlinked. Often, technical and economic indicators represent socio-cultural changes. But the purpose of monitoring is important: we should try to observe learning processes, changes in behaviour and attitudes.

This is why we suggest that impact monitoring should focus mainly on socio-cultural impacts.

What does this imply for an NGO?

Project proposals usually emphasize the technical and economic results aimed for. The socio-cultural impacts, although the main purpose, are often left out because they cannot be planned, and, moreover, are too difficult too measure or describe. But what happens if the project fails technically because the internal decision-making within the self-help groups (or within the NGO!) necessitated time-consuming and painful learning processes? Was the project then really a failure if all the individuals and organizations involved learned from it?

increase of self-esteem

If the learning processes are unknown, the contribution that self-help groups and NGO staffs make to the evolution of a project is underrated. Monitoring of socio-cultural impacts will not only bring recognition and esteem from outside, but also increase the self-esteem of the actors involved, both in the self-help groups and in the NGOs.

monitoring, a learning process

The PIM field phase has shown that the learning processes have not only been made visible by monitoring – **monitoring itself is a learning process!** For this reason in particular, it is not necessary to start with a complicated monitoring system: **start small, develop it by doing!**

start small, develop it by doing

Implications for impact monitoring

- don't use „impact" in a restrictive sense
- all subjectively important changes may be impacts
- different actors will observe different important changes
- plans focus on technical and economic impacts
- PIM focuses on socio-cultural impacts which are more relevant in the long term.

1.4 Participatory Monitoring

Participation

This is a wonderful-sounding term which is notoriously likely to be misunderstood, and is frequently no more than a cliché. In Part 1.1 we said that although the various categories of actors each have "their" projects, the main actors should be the people themselves.

Moreover, no management tool can be participatory in itself. Participation requires special attitudes and these participatory attitudes can be learned! Also, participatory monitoring is not participatory per se. The utilization of a tool is an art, and the tools have to be adapted to the conditions of the users. This will require appropriate methods.

In the context of self-help promotion, the word „participation" does not mean „to take part in a joint activity". It means more: there should be a **continuous empowerment of people's groups**, going hand in hand with a continuous relinquishment of power by NGO and funding agencies. Participation also implies an empowerment of the NGO vis-à-vis the funding agencies.

continuous empowerment

Participation is thus an ongoing process where one side learns to act increasingly autonomously, and the other side learns to hand over responsibilities and power.

> Participation is an **ongoing process** of capacity-building which requires **ongoing changes**.

Therefore, a participatory management concept serves to develop management tools which can be applied to increase people's independence of NGO and funding agencies, and NGO' independence of funding agencies.

Participation and monitoring

We have already stated that monitoring and evaluation aims to enhance people's ability to reflect and to learn. Participatory monitoring should help those involved to draw conclusions for decision-making out of this trial-and-error process, and to guide their activities according to „lessons learned".

Coming back now to our three types of organizations – the actors –, what do the organizations involved feel it is important to observe? If we let them answer freely, we see that there are many expectations and fears beyond the planning framework

- there are subjective interpretations
- there are all kinds of "hidden" expectations, and
- each actor also wants to observe the other actors.

own priorities

Whether consciously or unconsciously, each actor and each group of actors has his, her or its own priorities. Consequently, there are different autonomous monitoring and evaluation systems. They are usually not systematic or formalized.

These autonomous M+E systems of each actor must be preserved: they cover special aspects of the project process, and interlinked they will give a more complex view of reality than a single M+E system. To a certain extent, there should be information exchange and joint reflection – and this is the basis of participatory monitoring.

autonomy and dialogue

Accordingly, for participatory monitoring, participation means safeguarding and strengthening autonomy and establishing a dialogue between the actors as equals regarding their joint project.

Implications for participatory monitoring

- participation means a process of empowerment and increasing autonomy for previously disadvantaged groups
- each organization should have autonomy over its own monitoring
- mutual support and assistance among the actors is everybody's interest
- the findings should be periodically compared and sometimes reflected on jointly: this gives a more complete picture.

2. NGO-based Impact Monitoring

2.1 Advantages and obstacles

At the beginning of this booklet, we listed some conditions which should be fulfilled for PIM to be applied successfully. We then described the basic ideas underlying PIM. As we have seen above, there are advantages in theory, but perhaps obstacles in practice.

hurdles of PIM

PIM holds out the promise of being useful – but certain conditions must be fulfilled:

PIM is an appropriate tool for managing a self-help project –
   but do you really want to try a new monitoring concept?
PIM may take more time –
   but do your staff members want to spend more time on joint decisions?
PIM is intended to empower people who have no voice –
   but will the people in power at the moment accept a loss of influence?
PIM makes the project and the organization more transparent –
   but is this transparency really desired?
PIM may cause some conflicts if there are divergencies between attitudes, expectations and objectives. Are you ready to confront them?

Is your organization strong enough to tackle this new task? Are the leaders and members willing to face the hurdles mentioned above? If so, it is very likely that you will be rewarded with fruitful improvements of your activities!

reasons for PIM

There are several other good reasons why an NGO should adopt PIM:

– In an NGO there is a lot of **knowledge concerning socio-cultural impact lying fallow.** The staff have a large body of experience with learning processes in similar groups or projects. Field workers continuously observe the changes in their clients' environment. In order to improve

knowledge lying fallow

PIM 2 · NGO-based Impact Monitoring   19

project management, NGO-based impact monitoring could **mobilize and systematize this knowledge.**

value of field work
- In some organizations, the results of the **field staff's and social workers' endeavours are not perceived or appreciated**, especially when they are „invisible" rather than technical or economic. NGO-based impact monitoring could contribute to personnel guidance and team development by **demonstrating the value of field work.**

management of an entire organization
- **Non-profit organizations** do not have economic, but **intangible objectives**. Their performance is measured not by economic indicators (like profit or turnover) but primarily by social, cultural or other qualitative criteria. NGO-based impact monitoring can also be important in **managing the development of an entire organization.**

If there are conflicts, open or hidden, PIM helps to bring them to the surface. Misunderstandings can then be resolved if the actors are ready for open dialogue.

While NGO-based PIM is equivalent to group-based PIM, it is not merely a copy of it with different actors. The purpose of NGO-based PIM is to accompany group-based PIM and, to some extent, to complement it.

NGO-based impact monitoring consists of **three elements:**

A. Monitoring of socio-cultural impacts
is similar to group-based PIM (see booklet 1). The differences are:
- the field workers are the main actors, they select and specify indicators, observe, document and analyze the changes, and make (or prepare) decisions;
- monitoring focusses even more on socio-cultural impacts, i.e. learning processes, capacity building, changes in behaviour.

B. Joint Reflection Workshops
are regular joint meetings of the NGO and the group. Results of group-based impact monitoring are compared with NGO-based monitoring of socio-cultural impacts.

C. Facilitation of the PIM process
and ongoing accompaniment of the group by the development organization or NGO are necessary for the introduction and functioning of PIM.

Introducing PIM to NGO field staff

Before introducing PIM in a self-help group, it is crucial that an in-depth introduction to the PIM concept be given to the NGO staff involved, as they will be part and parcel of the implementation. This introduction to PIM, however, should be initiated carefully, starting with the field workers' practical experience.

Therefore, the introduction to PIM **should not start with a theoretical explanation** of monitoring or impact. The field staff should be encouraged to bring in their own experience and ideas. It must be made clear that their experience is valued highly – so start with questions:

start with questions

- What important changes for the people has your work induced?
- Which changes are normally reported on? Which changes are often ignored?
- What has changed in people's behaviour? What have they learned?
- Have other groups learned from these experiences?
- Is it possible to find simple indicators for these changes?
- How far were these indicators observed by the group members?

PIM is introduced through workshops. Three sessions will probably be necessary to sensitize the staff and explain the concept. Methodologically, the introduction could contain the following:

1) field staff's observations, ideas and experience on monitoring, impact, participation
2) short introductory speech
3) short handouts using visualization, pictures, examples
4) detailed discussions on advantages and risks of PIM to field staff and group
5) adaptation of the proposed concept to the actual work of the field staff

Only after the field workers' concrete examples of socio-cultural impact have been collected and analyzed should a more theoretical introduction to PIM be given. It will be very important to listen to and respect the field staff's comments concerning

theoretical introduction

- the viability of the concept,
- the additional workload, and
- the usefulness of impact monitoring.

> PIM does not ask for scientific definitions, but for subjectively important changes. The various actors have different views which need not be contradictory, but which are interlinked and have to be compared.

PIM 2 · NGO-based Impact Monitoring

2.2 Steps in introducing and carrying out NGO-based monitoring of socio-cultural impacts

As mentioned above, the procedure for NGO-based monitoring of socio-cultural impacts is similar to that of group-based impact monitoring, which was described in Booklet 1. The following description is therefore merely a brief outline.

Preliminary Step: What do we know about the context?

Certain essential information about the situational context should be available before PIM is introduced. It is then easier to adapt PIM to specific needs and integrate it in a given context.

Apart from this, if possible, you should as a rule try to use participatory methods for situation analysis which also serve for planning, monitoring and evaluation, such as

- PRA: **P**articipatory **R**apid **A**ppraisal
- PAR: **P**articipatory **A**ction **R**esearch
- SWOT: **S**trengths, **W**eaknesses, **O**pportunities, **T**hreats
- PALM: **P**articipatory **L**earning **M**ethods
- GRAAP: See, Reflect, Act (with the help of pictures)

These methods are based on ideas similar to PIM. You should make use of these to permit a realistic assessment of how people see their situation, their problems and needs.[1]

---

[1] For further details, refer to the GTZ publication *Participatory Learning Approaches in Development Cooperation: Rapid Rural Appraisal, Participatory Appraisal.*

STEP BY STEP

Steps in introducing PIM

1.	**What** should be watched?	expectations and fears of the staff members with regard to socio-cultural changes
2.	**How** can it be watched?	concrete examples of how these changes can be observed (indicators)
3.	**Who** should watch?	elected staff members who are directly involved in the respective project
4.	**How** can the results be documented?	records, tables, graphs, charts, descriptions

Steps in carrying out PIM

5.	**What** did we observe?	reports at the beginning of staff meetings
6.	**Why** do we have these results?	assessment and analysis by the staff
7.	**What** should we do?	immediate decision (or preparation for a decision) at the meeting (= adjustment of plan)

Step 1: What should be watched?

The project team like the self-help group should make a note of some of the expectations and fears concerning the self-help project. As far as possible, these should relate to the socio-cultural impacts: skills and learning processes in the group. In this context the following questions are helpful:

expectations and fears

– Based on your experience, what socio-cultural changes do you expect or fear from this project?

skills and learning processes
- What socio-cultural impacts resulted from similar projects?
- Which impacts were barely registered by conventional monitoring instruments?
- On the basis of which factors were they clearly recognizable?
- Which socio-cultural indicators should we be aware of in order to manage this project?

In this way, some of the hypotheses on the future development will be identified. It makes sense to discuss these hypotheses and indicators with selected resource persons from the group.

This procedure may lead to a result which is identical to the objectives of the NGO, or to the overall goal, project goal and expected results of the project. If they are fully congruent, so much the better. Whereas in conventional monitoring we rely on the formal information of the planning documents, PIM relies more on an informal assessment by the field workers and NGO staff.

example

In the case of the store of the Housewives' Committees in Caracoles, Bolivia, staff members had a number of expectations and fears. The prioritized aspects included the following:

EXPECTATIONS	FEARS / DOUBTS
that prices would be lowered and better quality offered than in other shops	*Should they sell for cash only or give credit?*
that the Housewives' Committees would take over responsibilities	*that they would not be able to recover the credits*
that a higher level of integration of the three Caracoles cooperatives would be achieved	*that the cooperatives might not pay the store*
that an opportunity would be created for women to participate in the cooperatives	*that the women might not be able to administer the store themselves*

Step 2: How can it be watched?

„Monitoring" and „indicators" are often quite abstract terms. Those responsible for keeping a watch should be encouraged to report very simply, on the basis of their experience, how they can tell that people have learned and changed. The procedure for mobilizing the practitioners experience should be similar to group-based impact monitoring. Concrete examples of how the social environment changes can be presented by each team member.

simple, concrete examples

The reasons for these changes can then be analyzed and the most vivid examples chosen for illustrative purposes. In accordance with what was said concerning group-based monitoring, there is no fixed procedure for deriving indicators from these examples. It is not a problem if no measurable or scalable indicators can be found, because in NGO-based impact monitoring descriptive examples can be observed and documented as well.

The expectations and fears to be observed may be chosen in Step 1, or definitively decided upon here in Step 2.

In view of the work burdens project teams are faced with, it is not necessary initially to introduce more than three to five indicators for a project. The necessary number of indicators, however, will depend on the complexity of the organization and its activities and hence its monitoring system.

PIM is easy to link to a conventional monitoring system. If the NGO has one, it must determine how far the socio-cultural indicators should be integrated into it.

example

In Caracoles, the NGO staff selected the following indicators for observation:

EXPECTATIONS/FEARS	INDICATORS
that prices would be lowered and better quality offered than in other shops	the prices of 20 staples in the shop are below prices in other shops nearby
that they would not be able to recover the credits	credits given each month do not exceed the cash payments received
Should they sell for cash only or give credit?	
that the cooperatives might not pay the store	
that a higher level of integration of the three Caracoles cooperatives would be achieved	the cooperative leaders held monthly meetings to analyze results
that the Housewives' Committees would take over responsibilities	the committees were able to check the discount efficiently

INDICATORS (derived from expectations or fears)	OBSERVATION METHODS
the prices of 20 staples in the shop are below prices in other shops nearby	list the prices of each staple monthly for the shop and for 5 other shops nearby separately; reply YES or NO and note down what was observed and what comments were made.
credits given each month do not exceed the cash payments received	list credits granted and payments received; reply YES or NO and note down what was observed and what comments were made
the cooperative leaders held monthly meetings to analyze results	reply YES or NO and note down what was observed and what comments were made
the committees were able to check the discount efficiently	price lists and prices charged are checked monthly by the committees; reply YES or NO and note down what was observed and what comments were made

Step 3: Who should watch?

The socio-cultural impacts chosen as indicators are frequently those which have already been observed by the NGO personnel. It is therefore best for the field workers and other NGO staff members to observe the selected changes themselves. One or two people should be chosen to be responsible for observation.

field workers or staff members themselves

However, it is also useful to confirm one's own views by cooperating with other people or organizations who know the project environment: a teacher, a priest, or staff members of other NGOs, or any insider concerning the group's internal structures. Whether this is appropriate will depend very much on the specific conditions.

confirmation of own views by co-operation

While it is not the purpose of NGO-based PIM to employ members of self-help groups as observers, they need not be excluded. It is first and foremost the field staffs' view which is of interest here. The views of the self-help group and NGO will be compared later in the joint reflection workshops this is part of the learning process!

PIM 2 · NGO-based Impact Monitoring

Step 4: How can the results be documented?

records, logbooks

As explained for group-based impact monitoring, a record must be kept of the impact observed. If it is done in the same way as for conventional monitoring, it will be recorded in a kind of logbook.

example

In the Caracoles project, monitoring forms were developed for noting down the prices observed in the various shops each month. For other indicators, there are questions which have to be answered with yes or no, with a blank space for remarks and comments.

Graphs and charts are also helpful for visualizing quantitative indicators. Indicator no. 2, about credits granted and payments received, was drawn as a bar chart.

Which information and for whom? When and how?

The NGO has to set priorities in information flow in accordance with the decision-making structures. This means that not all information has to flow to the NGO director or even to the funding agency, only summaries from time to time, or to report outstanding successes or conspicuous failures.

confidential treatment, clear rules

These rules for information flow should be worked out jointly by the entire NGO team. They should also decide what kind of monitoring information is regarded as sensitive or confidential, and establish clear rules as to how it should be handled: who must be excluded from the information flow?

In addition, a decision must be taken on how information should be fed back to the group. In Caracoles, the women were informed monthly. A minimum would be reports at the joint reflection workshop.

Important: All those involved must accept that although each organization wants to know what the others are doing behind the scenes, every actor is entitled to confidential treatment of his inside information. Not everything has to be analyzed: some secrets are best left under wraps!

Steps 5 to 7: What did we observe? Why? What should we do?

The steps to answer these questions are now basically similar to the steps described in Booklet 1 (Group-based Impact Monitoring).

Also, an NGO might already have a (conventional) monitoring system. Socio-cultural impacts might be monitored in a similar way. PIM is compatible with general management rules.

As mentioned in Section 1.2, the depths and periodicities of reflection and decision-making are different in every organization: you should match your PIM rules to your project and organization structure.

2.3 Joint Reflection Workshops

PIM comprises different autonomous monitoring schemes of various actors (self-help group, NGO, FA or other organizations/groups) in a single project. The individual actors observe the area that interests them most.

Advantages:
- it is not necessary for everyone to gather all the data, and the amount of data stays manageable for each actor;
- the facts are seen from different perspectives;
- more information is available for joint decision-making.

These different perspectives complement each other, and can portray the project reality more completely and realistically than a single actor could. To this end, the observations of those involved must be communicated and discussed regularly. The Joint Reflection Workshops fulfil this important need.

They provide a forum for exchanging and evaluating information. The various actors also hold a mirror up to each other, enabling them to compare the way they see themselves with the way others see them.

participants

The Joint Reflection Workshops are held regularly together with the self-help groups. They will be organized by the NGO to reflect on the progress of the project, though not so much in the sense of the planned „project outputs" as in the sense of impact assessment. These workshops should be arranged in a manner and an environment that the self-help group is familiar with. The participants are the parties involved (for ex.):

- self-help group
- NGO field staff
- NGO office staff

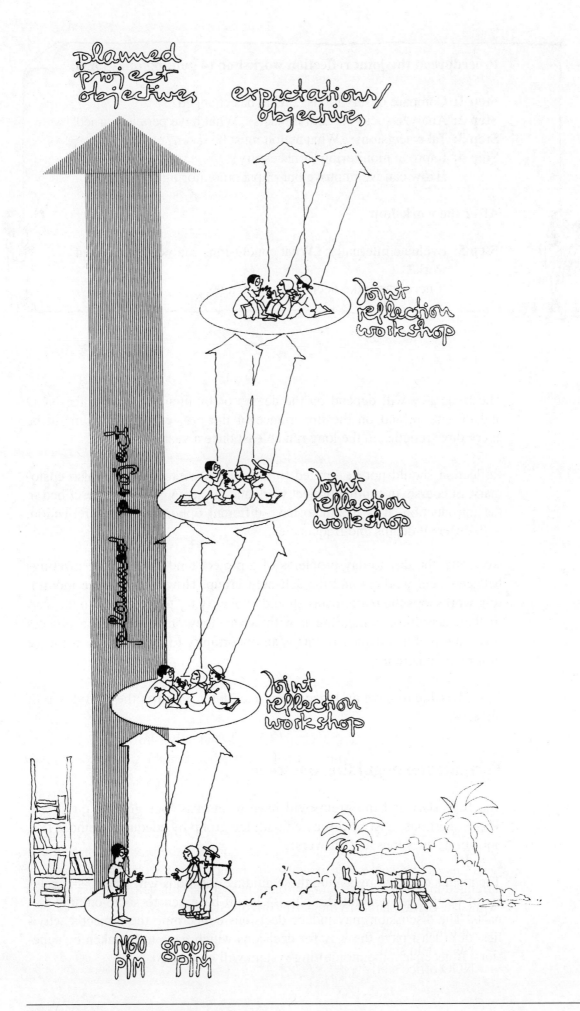

> **Procedure at the joint reflection workshop (4 guiding questions):**
>
> **Step 1:** Compare observations: „What has changed?"
> **Step 2:** Analyze socio-cultural impacts: „What have people learned?"
> **Step 3:** Take decisions: „What action must be taken?"
> **Step 4:** Improve monitoring (if necessary):
>   „How can we improve our impact monitoring?"
>
> **After the workshop:**
>
> **Step 5:** Evaluate internally: „What conclusions can we draw for our work?"
> If necessary take decisions on changes.

frequency — The frequency will depend on the degree of familiarity between the NGO and the group, and on the importance of the project. At first it might be every three months, in the long run at least once a year.

reflection — Reflection should not be limited to these rare workshops; it is also customary, of course, in routine project work and at meetings. But as described in the introduction, it is useful to have different opportunities for reflection, with different depths and frequencies.

looking back, comparing — Normally, the day-to-day problems of a project tend to dominate meetings between field workers and the self-help group. However, at these monitoring workshops the participants should explicitly try to look back to the start of their activities, comparing it with where they stand now. In a process involving prolonged activity this is an opportunity to stop for a moment for profound reflection.

It is advisable to have an independent facilitator who has the confidence of all sides.

Formulation of guiding questions

The field staff and the group will have to discuss four guiding questions. These questions must be discussed with the group by adequate methods (i.e. not simple questions and answers).

open questions — The results obtained by asking these guiding questions will lead to an analysis of the project context. Emphasis should be given to socio-cultural impacts. The discussion may induce decisions concerning the project's activities, or at least pave the way for decisions which have to be taken by superiors. If necessary, the monitoring system will be revised.

The questions must be open; the NGO should not influence the replies. Although NGO staff and field workers might have their answers to the questions, it is still important that they should first ask the group, by appropriate methods, and only afterwards encourage discussion by introducing their own observations (if necessary).

group's answers first

Workshop Step 1: What has changed?

The monitoring workshops start with (Step 1) the general guiding question **„What has changed?"**.

This is to compare the results of the group-based monitoring system with the the results of the NGO-based monitoring system.

Comparison of observations

The general question **„What has changed?"** leads to some deeper questions. Some questions refer to **change:**

- What/Who has changed?
  (This question is meant to introduce the report on the group's findings, which are a result of the group-based monitoring system.)
- What has caused the change?
  (the individual members, the NGO, or other factors?)
- How has it changed?
- How has this change affected you?
- What other change(s) has/have occurred as a result?

Clearly, these questions cannot be limited to socio-cultural impact; they include everything that is important to the group. The discussion should nevertheless focus on socio-cultural impact.

PIM 2 · NGO-based Impact Monitoring

Workshop Step 2: What have people learned?

various learning processes

When analyzing the socio-cultural impact, the NGO personnel not only have to refer to the indicators formulated in their monitoring system. They should try to grasp the various learning processes as a whole, and this is perhaps more feasible by asking open questions than by a strict comparison of isolated indicators. NGO members should be aware of the discussion needs of the group and use the opportunity of the joint reflection workshop for an open dialogue with the group.

What have people learned?

- Have the members of the group taken on new responsibilities?
- How far have the group's internal and external relationships changed?
- How far has the internal structure of the group changed?
- What new activities have been started by the group (or by members of the group)?
- What similar activities have other groups (or individuals) started?

Workshop Step 3: What action must be taken?

decision making

The next step is decision–making. The analysis of the findings will be aimed at achieving unequivocal results here. In keeping with the importance and frequency of the Joint Reflection Workshop, the decisions taken here tend to be of a strategic nature. That is to say, they indicate the basic direction and provide a framework for the solution. Operational decisions should be taken subsequently, at other meetings.

What action must be taken?

- What should the members of the group do?
- What should the project team do?
- What should the other people involved do?
- Who else should be brought in?

Workshop Step 4: How can we improve our impact monitoring?

The last step in the workshop is fairly general: if important issues have previously been neglected the monitoring system must be revised. In such cases it is useful for all those involved to agree at the workshop that these issues should be included in the monitoring system. Alternatively, each organization may take a decision at its own evaluation meeting after the workshop.

Potential improvements in PIM

- Which criteria and indicators should be improved?
- Which criteria and indicators are no longer necessary?
- How could the observation and assessment system be improved?
- How did you feel in our reflection workshop?

The case studies from the PIM field phase showed that many expectations/fears and their indicators which had been identified in the first meetings tended to be of short-term interest. By revising the monitoring system periodically, aspects which are of long-term interest are automatically sifted – and thus relevant indicators for the sustainability of the project come to the fore.

Post-Workshop Step 5:
What conclusions can we draw for our work?

After the monitoring workshop, the NGO should internally evaluate the results of the NGO-based impact monitoring. This reflection should go beyond the management of the actual project: instead, it should refer to more fundamental questions which often relate to the development of your own organization. If changes are necessary, decisions should be taken immediately.

The following questions should be discussed in the NGO after the workshop:

With regard to project management:
- What have we achieved? How have we achieved it? Who has assisted us? What has helped us to achieve this?

- What conclusions can we draw from the comparison between the group's observations and our own?
- Does anything need to be changed in our activities?

With regard to PIM:
- Which socio-cultural indicators need to be taken into account in our own regular monitoring and evaluation?
- How well did the original proposal work?
  Has it been modified by the beneficiaries' opinions?
- Should a questionnaire be circulated at regular intervals to document the specific activities of the projects, beneficiary groups and the context?

With regard to your own NGO:
- What can we learn from this project that should also be considered in other projects?
- Should we introduce new internal rules to improve our cooperation?

Concluding remarks

feed back

For the field staff, it will be important to learn about people's perception of their work. The participation of an external facilitator will further reinforce this. The feedback is useful for the field worker's self-assessment. Joint analysis of observed changes by the people's group and the field staff will increase the appreciation of successes which were previously concealed or rated as „merely subjective" impressions.

2.4 Facilitating the PIM process

facilitator as catalyst

Generally, PIM will be introduced in a group or people's organization on the initiative of the NGO. To a certain extent, therefore, the promoters are responsible for ensuring that group-based impact monitoring works.

For the NGO staff, it should again be emphasized that all of the group-based PIM and most of the impact analysis at the Joint Reflection Workshop is done by the group. The facilitator acts as catalyst. S/he does not tell people how to interpret the results of observation, but if necessary s/he will guide the group by asking questions. For group-based impact monitoring, it is crucially important that the community (not the NGO's field staff) identifies with the analysis.

You should not risk nipping group-based impact monitoring in the bud due to disagreements. If the group's and the NGO's observations differ, other opportunities should be provided or created for analyzing the content and the causes (e.g., the post-workshop step). If the staff member cannot cope with the conflict of roles, i.e. acting both as facilitator and as NGO representative, other facilitators must be found.

How can PIM be introduced into the self-help group?

Especially at the beginning, it is uncertain whether the idea has been well explained and understood. How do you carry on after the expectations and fears have been identified? To find out what is important, to find observable indicators, to find good observers, to observe, to feed back the results to group meetings – all this is easily said but not so easily done.

The NGO staff involved in PIM will probably have to stay in the village or with the group for several days when introducing the basic steps for the first time. During the first few weeks after setting up the first „draft" of group-based impact monitoring, the observers will need support. Presentation of the findings at group meetings should be facilitated by NGO personnel for the first year at least. How this is done is shown in examples from the field phase given in Booklet 3.

This may seem to be a lot of additional work; but if PIM is introduced when a new project is implemented, when we assume that frequent visits will be needed in any case, extra visits specifically for PIM will probably be unnecessary. And don't forget: PIM will definitely also help save time because

ACTION WITHOUT REFLECTION IS A WASTE OF TIME!

We need the cooperation of other practitioners and thinkers to test and to improve PIM.

If you are implementing and testing PIM in your project area we would be very interested to hear from you. Write and tell us about your experience with PIM.

We are planning to organize more regular and more efficient exchanges, if a substantial number of practitioners continue with the development and adaptation of PIM.

Please write to

FAKT	or	GTZ - GATE (ISAT)
Association for Appropriate Technologies		German Appropriate Technology Exchange
Gänsheidestraße 43		Postfach 5180
D - 70184 Stuttgart, Germany		D - 65726 Eschborn, Germany

Thank You

Eberhard Gohl / Dorsi Germann

Participatory Impact Monitoring

Booklet 4:
The Concept of Participatory
Impact Monitoring

A Publication of Deutsches Zentrum für Entwicklungstechnologien – GATE
A Division of the Deutsche Gesellschaft für Technische Zusammenarbeit (GTZ) GmbH

The authors:

Eberhard Gohl, economist and sociologist, spent a few years in Turkey, Peru and Bolivia. For eight years, he worked mainly with FAKT, DSE-ZEL and GTZ as a consultant for project management and organisation development. At present, he works as Controller in the German Protestant Churches' funding NGO „Bread for the World".

Dorsi Germann, sociologist and graphic artist, spent four years working in a community development project in Senegal. For the last fourteen years, she has been a consultant on adult education, appropriate technologies, technics of communication and visualization, project management, monitoring and evaluation, organizational development and participatory methods in Africa, Asia and Latin America, mainly working for GTZ and FAKT.

Die Deutsche Bibliothek – CIP-Einheitsaufnahme

Participatory impact monitoring : a publication of Deutsches Zentrum für Entwicklungstechnologien – GATE, a division of the Deutsche Gesellschaft für Technische Zusammenarbeit (GTZ) GmbH. – Braunschweig ; Wiesbaden : Vieweg.
  ISBN 3-528-02086-5
NE: Deutsches Zentrum für Entwicklungstechnologien <Eschborn>

Booklet 4. The concept of participatory impact monitoring
  / Eberhard Gohl/Dorsi Germann. – 1996
NE: Gohl, Eberhard

The author's opinion does not necessarily represent the view of the publisher.

All rights reserved
© Deutsche Gesellschaft für Technische Zusammenarbeit (GTZ) GmbH, Eschborn 1996

Published by Friedr. Vieweg & Sohn Verlagsgesellschaft mbH, Braunschweig/Wiesbaden

Vieweg is a subsidiary company of the Bertelsmann Professional Information.

Printed in Germany by Lengericher Handelsdruckerei, Lengerich

ISBN 3-528-02086-5

CONTENTS

1. Problem 3

2. Objectives of PIM 4

3. Situation analysis... 5
3.1 Management concepts for development projects 5
3.2 The project reality in self-help promotion.................................. 7
    From the expected continuity to the expected change 7
    From insisting on agreements to promoting the skills
    to solve problems 8
    From checks to acceptance of errors 8
    From evaluation to monitoring... 9
    From the factual level to the social level................................. 10
    From examining objectively verifiable impacts
    to perceiving the subjectively important changes 10
    From determining exact information to perceiving trends 11
    From formal to informal structures 12
    From project consensus to interaction between
    autonomous project actors.. 13

4. The structure of PIM 14
4.1 Combining the different approaches to solutions..................... 14
    Interaction of the project actors.. 14
    Valuation of the informal structures 18
    Perceiving trends 19
    Emphasizing the subjectively important changes 19
    Determining the socio-cultural impact..................................... 20
    Periodical reflection as monitoring 21
    Accompanying learning processes 22
4.2 The applicability of PIM 24
4.2.1 Invitation to handle PIM with fun and creativity 24
4.2.2 The monitoring steps.. 25
    Steps in introducing PIM.. 26
        1. Expectations and fears.. 26
        2. Indicators 28
        3. Observers.. 29
        4. Documentation 30
    Steps in implementing PIM 31
        5. Monitoring report 31
        6. Analysis 31
        7. Taking decisions 32
    Joint reflection workshops 32
    Facilitation of the PIM process 33
4.3 Prerequisites for and limits of PIM 33

1. Problem

„What's the point of development projects?"

We know very little about the impact of development projects, and even less about the social and cultural effects than about the frequently modest technical and economic successes.

Yet, critical inquiries regarding this aspect are arising more and more often. The funding agencies have to increasingly demonstrate successes to their financial backers, who are more often than not small donors or tax payers/voters. This pressure to be successful is passed on to the development organizations (NGO) in the South, which means that their staff have to increasingly prove the impact of their work. And the self-help groups or grassroots organizations for their part, albeit for the time being only those with greater awareness, start asking whether the NGO really needs so much money for their promotion or whether it would not make more sense for the funding agencies to support the self-help groups directly.

to prove success and impact

The funding agencies are adjusting to the new requirements. More and more often, impact analyses are drawn up and published. Yet, the limits of such analyses quickly become evident:

- The studies are conducted ex-post and barely affect the future of the project reviewed.
- The impact analyses are prepared by external persons and only include the standpoints of the relevant NGO and self-help groups to a limited extent.
- The studies are commissioned by a funding agency; therefore, at the most, they influence only future decisions at this level. They do not trigger any learning processes in the NGO or self-help groups who are actually responsible for implementing the project.

- The information that is published serves as justification and is not intended to promote learning. One is afraid of the public criticism that may arise if problems and errors become known.

The point is to find new solutions. To this end, a new **concept called Participatory Impact Monitoring (PIM)** was developed.

2. Objectives of PIM

The following task was defined for the PIM team at the beginning of the project:

The objective of PIM is to improve the realization of projects by

impact
- orienting the project along the socio-cultural impact;

autonomy
- promoting autonomous activities of the people;

cooperation
- improving the flexibility of and interaction between the development organization/NGO and self-help group.

self-help promotion

GATE (German Appropriate Technology Exchange, the unit within GTZ responsible for adapted technology) did not commission the study in connection with large technical projects, but aimed at small projects conducted within the framework of self-help promotion and appropriate technology.

learning

Such projects are primarily considered learning situations that are designed to trigger learning processes. Thus it becomes evident that if one wants a regulatory instrument for development projects that is based on the joint perception of impacts, one needs a concept which is not limited to being a regulatory instrument of the relief organization but can also be generally used as a management tool by all NGO and self-help groups.

However, PIM should be compatible with the goal-oriented project planning (ZOPP), the official planning procedure of GTZ, because many of the above-mentioned projects are part of larger programs that were planned with ZOPP.

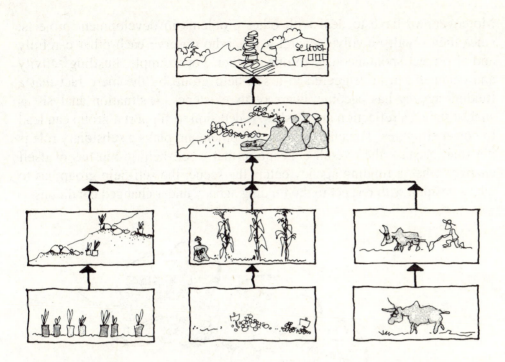

3. Situation Analysis

3.1 Management concepts for development projects

ZOPP is a self-contained, logical planning procedure from which a comprehensive project management system can be derived. ZOPP is a further development of the Logical Framework, which was developed for US AID. It seems suitable for all types of planning.

ZOPP

logical framework

When we plan the construction of a plane, all the marginal conditions are usually fixed and known: we can plan the course of proceeding quite precisely and only have to change our plans if something unforeseeable occurs, for example if a certain material does not meet our expectations.

A game of chess, a football game, or a development project cannot be planned entirely because the different participants react to one another. The marginal conditions are known to a limited extent only, and we expect to encounter new situations continuously. We cannot plan the course of events exactly unless we can exert great influence on the other actors. Instead we have to think over our manner of proceeding one step at a time.

It seems that ZOPP is a planning procedure more suitable for the tasks mentioned first, i.e. solving the problem on the basis of defined targets and conditions. This may be sensible in development policy on a macro level where many marginal conditions change relatively little and only infrequently.

Naturally, the contrasting representation given above is a simplification. In reality, the data used in connection with ZOPP have to be questioned frequently, and the management can be and is improved by the frequency and depth of the monitoring. But the premise remains that the framework is relatively constant and that we can foresee and influence it.

calculable

need for human- and process-oriented concepts

Moreover, we have to deal with other problems in development projects, since they usually involve several actors who observe each other carefully and also react spontaneously to one another. For example, bustling activity and changes can be triggered in a self-help group by the mere fact that a funding agency has become aware of its existence. A situation analysis as such stimulates reflection and wishes. A decision to support a group can lead to power struggles. The claim that development aid plays a subsidiary role is not true, because the overall conception that a self-help group has of itself changes when a funding agency enters the scene: the self-help group has to take a position with respect to the funding agency under changed conditions.

In self-help promotion, perhaps in development policy in general, we need new planning and management concepts that are based on the premise that everything is in a state of flux. We need such concepts at all levels: for the funding agencies as well as for the development and self-help groups. What is needed are concepts that take into consideration the continuous changes in the positions of all actors, i.e. concepts that are systematic and process-oriented.

3.2 The project reality in self-help promotion

In the colourful diversity of this world there are no magic recipes for correct project management. New concepts are continuously being tested, adapted and combined with other concepts in all areas of development cooperation. As in other fields, the management paradigms are always changing. A few trends which are already being implemented in self-help promotion and many other development projects will be discussed below.

no recipes

From the expected continuity to the expected change

> The only thing that is permanent is continuous change.

Deviating from a laid-out plan is normal. This is known by anyone who is familiar with the realities of projects at the grassroots level. Yet many funding agencies are attuned to continuity. Changes disrupt administrative processes and give rise to additional work. Development projects are supposed to change many things, but not themselves if possible.

revision of plans

Plans have to be revised not only because surrounding factors change, but also because the individuals involved in the project develop further – and that is, after all, an essential goal of development policy!

From insisting on agreements to promoting the skills to solve problems

> Conflicts and errors are normal!

continuous adaption and development

Not only the project activities, but also the agreements and even the organizations themselves, including funding agencies, have to have room to develop. This calls for continuous adaptation and development – not only of the methods but also of ways of thinking and attitudes:

Man:	Not merely carrying out, but analyzing and solving problems independently.
Power:	Not forced, but voluntary cooperation.
Success criteria:	Not purely technical and economic, but in harmony with socio-cultural aspects.

If we want to promote the ability to solve problems, i.e. the ability for creating innovations, and the efficiency of the people and organizations of the South, we cannot persist in adhering to old plans and agreements, but have to acknowledge changes, even if we do not like them. We must neither demand obedience nor fan the flames of rebellion; rather, we have to help the people concerned take action, so that they become actively involved and can identify with the results.

This cannot be achieved through administrative or formal measures alone. It also requires a positive, partner-oriented attitude that motivates responsible action. Then the quality of the project is not defined by the degree to which the plans were fulfilled, i.e. the largest possible congruence between planned and actual, but by the results that were actually attained.

From checks to acceptance of errors

> Intolerance towards errors is the enemy of creativity.

reflection

It is not without reason that many NGO in the South shrink from external evaluations: their experience with know-all evaluators has not been good. Nevertheless, evaluations always have two goals: control and reflection. Project evaluations should – especially if we view projects as learning processes – emphasize the second goal.

no unequivocal right or wrong

In the reality of projects there is hardly ever an unequivocal right or wrong. The future is so uncertain, the co-actors so flexible, reality so complex, perception so limited, and so on that we can only make imperfect decisions – indeed our decisions as individuals are even more imperfect than those taken in agreement with other actors. Yet, it is normal to make decisions that later turn out to have been wrong.

Errors are an important part of a learning process. It is no mistake to make errors in a learning process, if one is willing to learn from one's mistakes.[1] But this calls for a positive learning environment which permits self-realization of the individual and the group within the context of personality development, team and organization development.

learning from mistakes

From evaluation to monitoring

> He who comes too late is punished by life.

A classic differentiation between monitoring and evaluating states that monitoring is a process which systematically and critically observes the events connected to a project in order to control the activities and to adapt them to the conditions. Monitoring does not examine the planning. In contrast to this, evaluation involves an extensive analysis of the course of the project with the aim of adapting the strategy and planning to the conditions, i.e. evaluation does examine the planning.

continuous checking and adapting

---

[1] Naturally this should not be taken as encouragement to decide rashly or against serious risks. We always have to weigh the consequences of our actions and take responsibility for them. Then we will be able to deal with errors in an objective dialogue without emotional reproaches.

But this distinction is ideal-typical. In practice the borders cannot be clearly demarcated: there are different actors and decision-making levels in an organization or group, and the periodicity and depth of reflection – irrespective of monitoring or evaluation – will extend through several levels.

Many development organizations/NGO link their evaluation of the impact of a project to the infrequent external evaluations carried out by funding agencies. Others conduct evaluations more frequently on their own, for example once a year. If, however, we want to become aware of and promote learning processes by means of such regular reflective measures, we need a more regular rhythm.

From the factual level to the social level

> Development projects are implemented with people.

The logical reasoning of projects as a rule is that a project needs a concrete problem to be called into life. In other words, the project would then be defined as the solution to the problem, in which certain inputs would logically lead to determined outputs. In this way, projects are reduced to a technical-economic entity, and measurable, tangible success stands in the foreground.

attitude change

The socio-cultural component, however, can only be found in superior goals and marginal conditions, if it still fits into the prevalent logical system. And yet, this socio-cultural level is decisive: learning processes, changing attitudes and behavior – that is the development process we are striving towards, in the North and in the South.

From examining objectively verifiable impacts to perceiving the subjectively important changes

> The subjectively important changes have an impact.

Project applications often contain complicated, long-term and general objectives – the higher in the financing hierarchy, the more abstract they are. For the participants at the grass-roots level these general goals become less and less comprehensible or manageable. In that form they are far too abstract and not suitable as attainable objectives.

It is generally known that the further a problem lies in the far future and the less it affects a person's daily reality, the less the awareness of a problem. The more direct the effect of a problem, the more we are personally concerned about such problems. In this case, we are much more sensitive in our perception of signals of change and perceive the changes much earlier, albeit intuitively without being able to give logical reasons or proof.

The observations of all actors are defined by subjective perception and are often coincidental. Goal-oriented action, however, requires continuous, systematic and comprehensible detection of signals of change. Which changes are relevant with respect to activities is perceived most frequently and explicitly by those actors who are most affected by such changes. For us subjective evaluation is an important selection instrument, since it is easier and more effective than the external application of objectively verifiable indicators, which function mechanistically.

subjectivity as selection criteria

If we rely on the sensitiveness of the group, we will obtain finer and earlier reactions as well as adaptations to changes that – contrary to administrative decisions – give rise to early countermeasures.

From determining exact information to perceiving trends

Relying on intuition saves time.

Only phenomena that are based on absolutely precise information and lead to definite conclusions by means of Aristotelian logic are considered to be scientifically proven.

complexity

Yet, in actual fact, our world is so complex that we will never be able to definitely determine all factors and their interaction in our dynamic environment. In our daily lives we have to filter out the information that is important to us. Formal criteria help us make our decisions, but we are even more guided by our experience and intuition. We have to use our limited perception to recognize patterns and interpret them. In this way, we can obtain quite meaningful results and decisions with little effort.

intuition

As far as development policy is concerned, we would be unable to act if we were to conduct projects primarily on the basis of exact information and scientific conclusions. If we wanted to do justice to the complexity of the circumstances, we would have to collect piles of data – a practical limitation of many formalized systemic approaches. Then, however, the impression would be given that the protagonists, who are unskilled in formal logic, were not capable of helping themselves.

Monitoring systems that lay claim to a high scientific quality and feature a complicated language act as deterrents. Users fear that their observations will be unscientific and incorrect – or, to put it more succinctly, they are simply afraid of being incompetent. Thus, they hamper such systems.

indication of trends

In projects where self-help plays a significant part, we need simple monitoring systems and indicators that are easy to manage. Even though such indicators may not always be exact, they will illustrate essential trends quickly and plausibly. With such monitoring systems we will be able to manage self-help projects effectively.

From formal to the informal structures

> Not the formal, but the informal structures and communication are important.

We have to revise our thinking considerably both with respect to processes and changes as well as structures. If we know the project reality at the grass-roots level, we know that informal communication is normal and that the informal structures are the decisive ones.

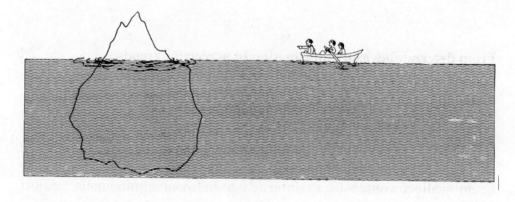

This point is best illustrated by comparing an iceberg to the highly formalized structures of the established organizations: although about 6/7 of the volume is hidden from view under the surface, it is of substantial importance for the colossus. The visible 1/7 is a structure that developed over a long period of time, i.e. an expression of the internal arrangement and the history of the organization. In young organizations, however, the formal structure frequently is merely put on, and it does not fit the internal processes. Thus, it is dysfunctional and only of slight significance.

Naturally project structures have to be formalized as agreements between different actors. But we don't need palaces, we need tents. We don't need organizations that are characterized by mechanistic-bureaucratic structures, because with time these lose their ability to perceive and to react properly to changes in the environment. We need flexible and innovative organizations, because they will be more efficient in connection with dynamic environments.

From project consensus to interaction between autonomous project actors

> Every actor has „his" project.

The self-help group has „its" project; the staff of a development organization/NGO have „their" project; the desk officer of a funding agency has „his" project; the consultant has „her" project. And quite frequently they all believe it is one and the same project. Within the framework of ZOPP, „the" project is planned on the basis of agreement, i.e. there is „the" project team, and „the project" sometimes has many employees and vehicles at its disposal.

different perceptions and interests

PIM 4 · The Concept of Participatory Impact Monitoring

Actually a project is a bundle of measures designed to solve a specific problem. In the field of development cooperation, the word „project" has a more extensive meaning – frequently it is characterized by the illusion that there is a common feature. The mutual project often is pure fiction – at the most there is a point of overlapping, a smallest common denominator, between the individual viewpoints and projects of the different actors.

interaction and coordination

This is so because each actor defines the problem he is confronted with differently, perceives it differently, and, thus, also defines his bundle of solutions differently. Each of these projects is different, and the cooperation between the actors is limited to the area where these different definitions overlap. Furthermore, each of these projects is planned, conducted, monitored, and evaluated by the respective actors – albeit in coordination with the others – and the responsibility is borne by each actor or by each group alone.

4. The Structure of PIM

The rules for applying PIM were already explained in Booklet 1 (for self-help groups) and Booklet 2 (for development organizations/NGO), practical examples were provided in Booklet 3. Therefore, the mode of operation of PIM will not be described in this booklet. Instead we want to point out certain characteristics from theoretical and practical standpoints.

4.1 Combining the different approaches to solutions

PIM tries to take up the approaches described above and to combine them in a new way. The following features are of particular importance in connection with PIM:

Interaction between the project actors

several autonomous strings

PIM involves several autonomous strings: the self-help groups observe what is important for the members of their group, it collects data and makes decisions. The development organization/NGO does the same on its own at first. The principle can be expanded at will: also the funding agency could proceed in this way and other groups. A federation of cooperatives could have separate monitoring systems for federation leaders and for the employed experts. The self-help group can differentiate by social groups or partial projects.

cooperation

Each group of actors covers its areas of interest; thus, it collects only a limited amount of data. If the groups were to regularly exchange their perceptions and interpretations, much broader coverage could be attained. The information would not be summed up mechanistically, but would form an overall picture due to the interaction between the project actors. A systemic mode of observation is not achieved by accumulating data, but only through cooperation between the actors.

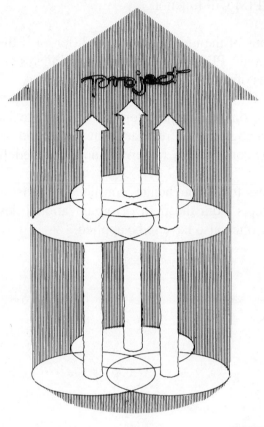

This concept of autonomous actors, whose areas of interest overlap at certain points, is a brief description of the meaning of participation within the framework of PIM. Participation is not understood to mean participating in a project managed by an outside party; rather participation is the responsible execution of one's own project in cooperation with other actors.

The project data and autonomous monitoring systems of the individual actors are discussed regularly and continuously at Joint Reflection Workshops. This enables the observations of all those involved to be utilized and considered with regard to the joint project goals.

The more congruent the aims and expectations of the individual actors are, and the more they are in agreement with the overall project goals, the more smoothly and efficiently PIM will function.

If the aims and perceptions of the individual actors differ, or if they deviate from the overall project goals, PIM acts as an early warning system, because discussion and the process of reaching agreement are disrupted.

PIM can have a corrective function if the actors are sensitive to such signals and are in a favour of change. In the regular Joint Reflection Workshops, aims and strategies can be continuously reviewed and jointly redefined.

If there is little willingness to accept change, conflict and division may occur. In such cases, concepts from the field of organizational development and conflict management often also have to be applied.

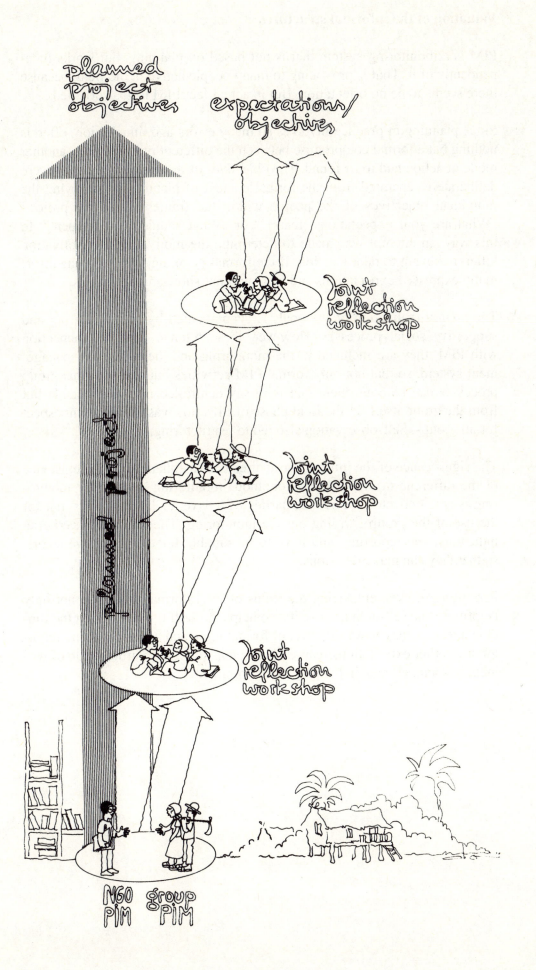

Valuation of the informal structures

PIM is a monitoring system that is not based on planning, but works independently of it. That is provoking to many people active in the field, because there seems to be no orientation. But in actual fact PIM is goal-oriented.

Since planning in practice involves so many errors and, in addition, often is nothing but a formal compromise between the different project actors, another mode of action had to be found. The objectives of the people concerned were deliberately separated from the formal context of planning by querying the individual objectives of the people within the framework of the project: „What are your expectations (fears)?" or „What should (not) happen?" In this way, an attempt was made to determine the motivation (or motivation killers) relevant to taking action. If the formal planning was good, the informally expressed expectations and fears will not show any deviations.

The *factual* realization of the goals is accompanied by both hindering and supportive social processes. However, these often escape. In connection with PIM, they are included in the monitoring and, hence, in the management system, so that not only formal PIM activities but also informal group processes can be controlled, that is to say controlled not from outside but from the group itself. In the field phase of PIM this was an important aspect for all groups. Self-observation also led to restructuring.

significance of informal processes and structures

The significance of the informal structures is also enhanced by the existence of the different strings where autonomous monitoring occurs: the monitoring is not conducted for outside parties who have unclear demands, but for the use of the group carrying out the monitoring. The observation criteria, indicators, and reporting only have to be suitable for the respective actors, so that they can make decisions.

Permitting or even enhancing the status of the informal aspects is not only helpful for the self-help groups, it also helps the staff of NGO. Even the funding agencies may leave the level of formal logic and agreements and monitor such other expectations/fears as leadership structure, participation of women, or social climate in the project environment.

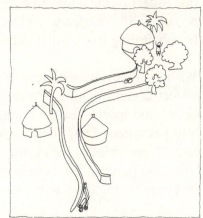

Perceiving trends

The more intangible the goals, the less exact the information that can be obtained. The aim of PIM is to encourage the actors to form hypotheses about their perceptions. Inaccurate observations are permissible. No formal indicators are expected. Even if the information is not accurate, it is first assessed within a group, verified or disputed, and, if necessary, supported by additional perceptions. The information can also be exchanged and compared with other project actors. Thus, the group serves as a filter and corrective instance.

group as corrective

PIM is by no means devoid of logic. It contains elements of both formal logic and networked logic („rock logic" versus „water logic" according to Edward de Bono). Within every PIM system, the relationship between expectations/fears, indicators and describability is questioned in formal-logical terms. However, there exist relations of networked logic between individual expectations and the different strands of PIM, which are defined by relevance of action, interests and subjective perception. Due to this, PIM – in view of its simplicity – is better able to do justice to a complex reality than a complicated but one-sided monitoring system.

In this way, new trends can be perceived at a relatively early stage, not only by individuals, who can be wrong, but they are analyzed collectively, and they can be quickly integrated in the group's decisions.

Emphasizing the subjectively important changes

PIM wants a solution that is subjectively the best for the actors. Contrary to conventional monitoring systems, PIM is not only oriented towards abstract project goals. These abstract long-term goals are frequently irrelevant with respect to the actual activities and problems of the actors. In view of the fact that PIM builds upon the fears and expectations of the members of the group with respect to the achievement of goals in the long-term, it concentrates on goals that can be achieved quickly and, thus, the activities have an immediate relevance.

short-term goals

This aspect of the subjective relevance of activities makes PIM attractive to the actors. In contrast to collecting data in order to attain an objective goal, the monitoring is regarded as being meaningful.

subjectivity

By the way, the special feature of PIM is not that it works on the basis of subjective information. Perception and evaluation are always subjective. Rather, the special feature seems to be that this kind of information is explicitly permitted. Thus there is no pressure to prove anything to an outside party. Nonetheless, during the field phase, there were always attempts to make the perceptions verifiable „intersubjectively" (= objectively) within the group during the internal discussions of a group of actors.

Determining the socio-cultural impact

socio-cultural impact

PIM is designed to monitor the impact on the project environment. Conventional planning – like monitoring and evaluation – concentrates on technical and economic effects. PIM wants to focus on the socio-cultural impact.

Yet, this seems to involve some contradictions: on the one hand, the expectations and fears are focussed on goals to be achieved in the short-term – and these usually are technical and economic; on the other hand, PIM considers the actors to be autonomous. So how can one tell them what they are to concentrate on?

For development organizations/NGO and funding agencies the following applies: Since PIM usually is supposed to supplement a conventional monitoring and evaluation system, it is not difficult to systematically obtain socio-cultural indicators.

The monitoring of the socio-cultural impact without any external specifications also worked quite well with self-help groups, as a matter of fact, much better than expected by the PIM team. The mechanisms:

- Many expectations and fears of people are connected to their abilities:
  „that we can manage it properly",
  „that the storehouse is refilled regularly",
  „that the bookkeeping works",
  „that we have more time for our families",
  „that all members participate", and so on.

- The people concerned often take technical or economic indicators as indicators for socio-cultural changes:
  „sugar content of the palm juice" = „that the members do not dilute the palm juice delivered with water any more";
  „the amount of quinua sold" = „the members' diets are healthier";
  „status of loans" = „that the members' payment habits have improved".
  What these expectations express in general is that „the members are acting more responsibly".

- Changes in the indicators are also very informative. On the one hand, they show to what extent the perception of problems is changing (changes in an indicator) or whether a problem is no longer considered important or has been solved (omission of an indicator). On the other hand, it shows whether these are short-term problems or a long-term process: the short-term problems disappear after a while; the long-term problems remain part of the monitoring system. Thus, the impact is also determined through the system's self-regulation.

- The application of PIM itself triggered many learning processes for persons and, in particular, in self-help groups, which were then observed by the staff of the development organizations/NGO (see Booklet 3).

PIM as a learning process

Let us make one more comment with respect to impact: First, PIM does not make a strict differentiation between changes, effects and impact. Rather it tries to identify subjectively important changes at the beginning. Only in a second step, does it determine how these changes are related to the activities of project actors and, thus, are effects. Thirdly, the permanence and range of the changes is determined by regular monitoring (see Section 4.2.2). Due to this self-cleaning mechanism, effects and permanence are filtered out automatically.

Periodical reflection as monitoring

In simplified terms, one could say that the project cycle is divided into situation analysis, planning, monitoring, and evaluation. The participatory approaches to project management always focus on one of these areas, most frequently on situation analysis, planning, and evaluation. Concepts such as PIM, which purposely concentrate on monitoring, are rare.

Even though PIM does not differentiate strictly between evaluation and monitoring, the term „monitoring" is purposely used in the name. The term „monitoring" is better suited to underscore the frequency and continuity of the perception, reflection and, if necessary, decision-making, which accompany the continuous activities.

process-oriented project-cycle management

PIM can be applied in every phase of the project cycle, since it is not based on formal specifications or plans. Thus, PIM is a tool for process-oriented project-cycle management. As shown by the examples, PIM also improves the management of the entire organization.

The key element is regular observation and reflection – at different intervals and to a different depth at the individual levels – within the group of project actors and between the groups. Vice versa, it can be concluded that the message of PIM essentially is as follows: People, open your eyes, think things over, sit down together and talk to each other!

Accompanying learning processes

reflection – a motor of change

Projects are learning situations, and learning situations are characterized by the intention with which the learners change. Indeed, every human action is a form of changing oneself if it involves reflection.

need for positive and informative feed back

Learning in learning situations should make people more capable of acting with respect to things and other people. It should help people develop special skills and a general ability to solve problems. In order to help the learners develop this ability to act, the feedback in learning situations has to be controlled. The learner needs room and has to be able to count on informative and non-punitive feedback; he must be able to make mistakes and improve by learning from them.

organization development

PIM also wants to fulfil these educational requirements. As described above and in Booklet 3, PIM accompanies people and organizations in their processes of change. In this respect, it closely resembles the concept of organization development.

It would be wrong to expect linearity in these learning situations, instead we expect irregular changes. We have to avoid using planning – knowingly or unknowingly – as an instrument of power. That obstructs any autonomous changes.

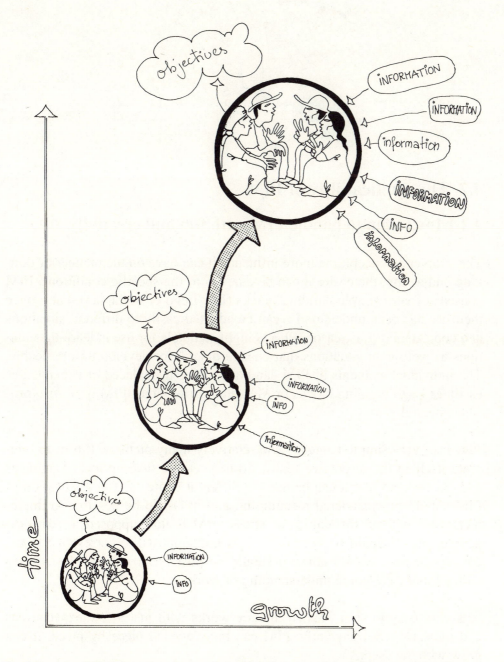

In this way, PIM does justice to its objectives, and it achieves more than a conventional monitoring system: PIM not only monitors the socio-cultural impact in the project environment through the cooperation of the parties involved, it also stimulates the parties, changes the behavior and attitudes of people and organizations, initiates and reinforces new learning. PIM makes an impact.

4.2 The applicability of PIM

4.2.1 Invitation to handle PIM with fun and creativity

simple and adaptable concept

The simpler a concept, the more influence it can have on the manner of thinking and acting, the easier it can be adapted to the specific conditions. PIM consists of very simple building blocks (like Lego or Metaplan) that – once their use has been understood – can be adapted easily to different situations and contexts. For this reason, PIM is highly suitable for use in learning situations as well as in situations that change quickly or develop unforeseeably. The individual elements of PIM can be rearranged, replaced or expanded as easily as Lego building blocks or Metaplan cards. Thus, PIM is *eo ipso* process-oriented.

fun and creativity

PIM uses very simple language and relatively many pictures. It tries to integrate itself in the respective reality, to respect and take up local languages and customs, so that it can be used in different contexts. The application of PIM should be considered meaningful, and it should be possible to implement PIM without difficulties or stress. PIM is not supposed to scare off people, and it should be fun to use. It is not supposed to correspond to the concept of hard work; rather, it should be possible to adapt it to cultures with a more easy-going understanding of work.

It is very easy to start PIM: It already works with just a few expectations and indicators. Subsequently, PIM can be expanded piece by piece; it can grow with the users.

different combinations

Every PIM system, no matter how small it may be, is coherent and independent. It monitors its own expectations, uses its own indicators, and results in its own analyses and decisions. If one part of the monitoring system is given up, other parts can be continued independently. Up until now we have group-based impact monitoring and NGO-based impact monitoring. Yet, other PIM systems can be integrated easily, not only for funding agencies, but also, for example, for supervisory boards of the development organizations/NGO or other committees of the self-help groups.

PIM can be combined with other concepts, i.e. not only with such that feature similar premises with respect to process-conformity and participation, but also with completely different concepts such as ZOPP. Thus, PIM can accompany and support the gradual discontinuance of an old concept.

4.2.2 The monitoring steps

Group-based impact monitoring can be integrated without any difficulties in the regular meetings of groups; thus, only relatively little additional time will be needed. Since the monitoring is supposed to be carried out by the members of the group themselves, the individual steps can be defined freely. The informal aspects should be enhanced and must not go under as a result of premature formalization.

In the same way, NGO-based impact monitoring can be integrated in team meetings. If these are not held regularly, PIM provides an occasion. As a rule, it should only be used as a supplement to conventional monitoring of the target structure, resources and frame of action. It is possible that indicators for socio-cultural changes are already being obtained with the conventional monitoring; otherwise one merely has to add the socio-cultural indicators obtained through the PIM concept.

easy integration in existing M+E systems

However, in those cases where conventional monitoring is not being conducted by the NGO, but the organization is interested in having such a system, PIM can also be used as a basic building block for a monitoring system. Due to its loose concept, it can allay people's fears and illustrate the feasibility of monitoring. Then it can grow slowly together with the NGO.

PIM 4 · The Concept of Participatory Impact Monitoring

The steps and elements of PIM are described and commented in detail in the booklets 1 and 2.

STEP BY STEP

Steps in introducing PIM

1. **What** should be watched?	expectations and fears of the staff members with regard to socio-cultural changes
2. **How** can it be watched?	concrete examples of how these changes can be observed (indicators)
3. **Who** should watch?	elected staff members who are directly involved in the respective project
4. **How** can the results be documented?	records, tables, graphs, charts, descriptions

Steps in carrying out PIM

5. **What** did we observe?	reports at the beginning of staff meetings
6. **Why** do we have these results?	assessment and analysis by the staff
7. **What** should we do?	immediate decision (or preparation for a decision) at the meeting (= adjustment of plan)

Steps in introducing PIM

---

1. **Expectations and fears:** at the beginning, the group determines which changes are most important for it:
   – What are our expectations?
   – What are our fears?

   Three to five important aspects are selected from the statements.

---

motivation

The introduction is based on the assumption that these expectations and fears are a significant motivation for the group's members to participate in the activities or in the self-help group. Constant observation of these subjectively important effects or changes leads to reflection and decision-making, and it is a means of self-controlling the project and, if applicable, the organization.

Naturally, the expectations and fears mentioned above are only the tip of the iceberg. Therefore, the concept provides that

- they be corrected and refined continuously;
- open questions investigate the actual, unforeseen changes;
- other significant expected or feared changes be included in the monitoring;
- a differentiation can be made between different subgroups (e.g. women, youth) that have different expectations and personal objectives;
- group meetings begin by asking about any observed changes, so that the group can take regulative decisions.

An important objection to the approach of querying the group's expectations is that it does not help define the „real" impact. Yet, what is the „real" impact? The further away the recipient of the monitoring report, the more abstract the concept of impact becomes. However, the significance of PIM is that it creates awareness of the fact that there are different actors with different points of view. A theoretical explanation of the terms „impact" and „monitoring" cannot be provided in an introductory meeting of a self-help group, and this is not necessary. The course of the process is marked by the corrective measures, e.g. whereas indicators for short-term problems will disappear after a few months, the significant problems will remain on the agenda for a longer period of time.

different points of view

corrective measures

> 2. **Indicators:** the group gives examples – as concrete as possible – of its expectations and fears. Simple indicators are formed on the basis of these examples.

starting simple

Anyone who criticizes the scientific quality and systematics of this type of monitoring has to accept that it is primarily concerned with learning. And learning always means starting at a simple level and moving on to more complex aspects: the indicators are formulated and developed further by the group.

During the field phase, many indicators had not been derived from the expectations or fears by means of formal logic. Possibly some indicators were chosen on the basis of other considerations, e.g. because they are observed by the people anyway or because they contain other expectations. By analyzing the perceptions repeatedly, the meaningfulness and quality of the indicators are examined. In a process of trial and error, the group can work towards better indicators.

improving indicators

Another striking factor was that relatively many measurable indicators were selected. Then the relationship with learning processes and behavioural changes is not obvious to outside persons and needs to be explained. Yet, as mentioned above, the relationship made sense to the group.

In addition, a relatively high percentage of the indicators required „Yes" or „No" as an answer; supplementary comments then described the background. At first sight such indicators may seem unsatisfactory, but they can be applied most easily by an unskilled group. Moreover, the comments, in which these indicators were made more precise, proved very informative.

> 3. **Observers:** the group designates observers („watchers") for the indicators/examples and defines how they are to report on these (e.g. on posters, at meetings).

The observers – formally – can be designated very quickly. Whether they then perform their tasks satisfactorily cannot be influenced from outside. Observers are expected to assume a high degree of responsibility and to work independently.

responsibility and independance

During the field phase, members holding senior positions were often chosen as observers. Thus, the group had influential observers who were obligated to assume more responsibility and provide transparency. At the same time, the creation of a parallel structure of power was avoided.

In many cases, however, it may be meaningful to confer the monitoring tasks on other persons from the group. Not only does this lead to more self-control within the group, it also tends to stimulate learning processes, responsibility, and commitment among the other members.

PIM 4 · The Concept of Participatory Impact Monitoring

> 4. **Documentation:** the indicators and their form, analyses and decisions are documented continuously.

This step is very important because with the documentation the organizations write their own history. It is a laborious task, but not excessively time-consuming. Many self-help groups and development organizations/NGO have already made it a habit to keep records of meetings. During the field phase, the groups started keeping monitoring booklets or documentation.

Nonetheless, this is a weak point in practice, because it requires individual persons to sit down and complete the records regularly. Then the documentation can also be used for outside support.

Steps in implementing PIM

> 5. **Monitoring report:** A short report is given at the beginning of every group meeting – prior to setting down the agenda:
>   a. How have the indicators changed? (This can lead to corrections and refinements of the indicators used until then).
>   b. What other important factors have changed? (This can help to decide whether additional indicators ought to be observed in future).

The manner of proceeding within PIM involves repeatedly monitoring specified criteria at regular intervals. In this respect it differs somewhat from many processes of self-evaluation and so on, where the situation analysis is based on open questions. Thus, PIM induces more regularity and, hence, shows the development of the group and its situation much more clearly.

regularity

Since PIM expects only limited continuity, however, there are inquiries with respect to additional changes. It remains unanswered at first what influences these changes, i.e. whether the changes are direct effects of the activities of one of the project actors. This is still of secondary importance in this step.

> 6. **Analysis:** if necessary, the group analyzes the origin of changes, for example on the basis of the following questions:
>   a. What did the persons involved (we ourselves/other project actors/external actors) contribute to the changes?
>   b. Which other effects/conclusions result from this?

responsibility

Now the actual analysis of the effects is conducted. It may be unsatisfactory in formal-logical terms, but the actors are encouraged to reflect on their responsibility.

A documentation of this effect analysis need not be drawn up. However, it can be prepared at any time, if necessary.

> 7. **Taking decisions:** after the analysis, the group defines its agenda and takes decisions.

PIM – i.e. regular analysis of changes – should form the basis for the decisions of the individual project actors. Thus, the decisions are based on factual reasons, and the members are enabled to participate responsibly. The leadership of the organization becomes more transparent and democratic.

Joint reflection workshops

When testing PIM this part was sometimes underrated. The joint reflection workshops are an important part of impact monitoring, because the individual monitoring systems are brought together here, i.e. the individual actors are confronted with the standpoint of the other parties involved in the project.

comparison of different viewpoints

This comparison of perceptions from different viewpoints underscores the project reality, provides more complex information on the changes, and gives rise to a deeper understanding of the impact of the project. PIM becomes a systemic approach without a formal-logical superstructure through the dynamics of the actors involved, since they continuously rearrange their positions with respect to one another.

Furthermore, this kind of exchange also affects the emotional level: the understanding of the project actors for one another is improved, and communication between them will be easier in future. By having a mirror held up to him, every actor can compare his self-image with the image of the other – and presumably can learn from this experience.

mutual understanding

Facilitation of the PIM process

PIM cannot be implemented overnight. The monitoring system needs to be introduced and needs to be accompanied at all levels until it functions automatically. For this purpose facilitators are needed, who support the group in its efforts to conduct an autonomous opinion-forming and decision-making process.

This is even more necessary in joint reflection workshops: for example, if the NGO does not lead the workshop well and slips into an attitude of self-justification, it risks blocking mutual learning.

4.3 Prerequisites for and limits of PIM

> PIM cannot be the solution to all problems.

PIM, first of all, should not be considered a set of instructions for activities, but a concept:

a concept

- it provides a simplified representation of a more complex process;
- it is limited to a manageable number of dynamic elements;
- it permits a planned and methodical manner of proceeding;
- it has to be reviewed and adapted to local conditions;
- it serves as an introduction to a learning process;
- it contains a temporary truth.

PIM is hardly formalized. When it is applied in practice to projects, PIM has to be arranged in many respects. When PIM has been defined and adapted to the respective conditions, it can be used as an instrument. Its value as a tool is also limited: The knowledge of how to use a tool is an art!

empowerment

PIM is not suitable for every project. It was developed for self-help projects where there are many processes and very few structures, where actors constantly react to one another, and where change is expected continuously.

adaption

PIM is not neutral with respect to interests. PIM is participatory. It affects power structures, since it works towards giving more decision-making power to previously disadvantaged people. Since PIM brings to light internal conflicts, changes decision-making mechanisms, and limits the influence of those previously in power, it necessarily provokes resistance.

Participation is desirable, but not at any price. It should not be overlooked that participation is practiced differently in different cultures. It can be dangerous to force – in a rush of well-meant missionary zeal – ill-prepared organizations or their members to accept and implement a western-style participatory leadership model. Not every organization can handle the transparency induced by PIM. PIM can work to break up authoritarian structures, but only if many members of the group want such changes and are willing to take responsibility.

The self-help groups should exhibit the following characteristics:

preconditions for organizations
- Internal participation, responsibility and decision-making power should be divided among several members to a certain degree.
- Continuity of the actors (slight fluctuation), internal consolidation.
- The task that is to be managed must be considered an important joint concern.

Prior to introducing PIM, the cultural background of the self-help group should be examined carefully to determine whether the following preconditions are met:

- Participation of women to a certain degree.
- Extensive ethnic and cultural tolerance.
- A certain degree of literacy.

The projects must have the following characteristics:

preconditions for projects
- Flexibility of those involved in the project, especially the NGO and the funding agency.
- Continuous and trusting relationship between self-help group and NGO.
- Mutual wish of the self-help group and NGO to control the project jointly by means of monitoring.

When it is applied regularly, PIM demands a relatively high degree of discipline from all parties involved. In particular the NGO feel the increase in the required amount of work when they did not plan to accompany projects that were initiated. Roughly speaking about 10% of the working time should in general be reserved for joint reflection, and the PIM concept can provide suitable guidance for this.

PIM also requires certain attitudes:

attitudes
- Mistakes happen! The fear of mistakes has to be reduced to make possible a certain degree of self-critical behavior and, thus, learning.

- All cooperation partners (for ex. self-help groups, NGOs, funding agencies) accept one another as competent partners who are willing to learn in their respective fields. Only then can autonomous monitoring and mutual comparison lead to a meaningful exchange, from which all project actors can benefit and learn.

- PIM calls for confidence in the ability of the partners and, thus, willingness to cooperate. This means that there must be no impatience or paternalistic behavior. It means that all parties involved must be willing to listen to one another, to let go of their own concepts in order to be receptive for the ideas of others and to be cooperative.

Although PIM is simple, it does require a certain amount of effort to teach its contents. The first organizations in which PIM was introduced did not understand the concept entirely. Here the introduction was accompanied by more intensive support. In one organization, however, where only a basic introduction to PIM was given, the PIM concept was quickly implemented. The successful communication of the concept of PIM still needs to be clarified.

PIM is a young concept which needs to be developed further. Right from its beginning it has been concieved together with partners from all over the world – in a process-orientated way.

We need the cooperation of other practitioners and thinkers to test and to improve PIM.

If you are implementing and testing PIM in your project area we would be very interested to hear from you. Write and tell us about your experience with PIM.

We are planning to organize more regular and more efficient exchanges, if a

PIM 4 · The Concept of Participatory Impact Monitoring

substantial number of practitioners continue with the development and adaptation of PIM.

Please write to

FAKT or	**GTZ - GATE (ISAT)**
Association for Appropriate Technologies	German Appropriate Technology Exchange
Gänsheidestraße 43	Postfach 5180
D - 70184 Stuttgart, Germany	D - 65726 Eschborn, Germany

Thank You

Burkhard Schwarz / Eberhard Gohl / Dorsi Germann

Participatory Impact Monitoring

Booklet 3:
Application Examples

A Publication of Deutsches Zentrum für Entwicklungstechnologien – GATE
A Division of the Deutsche Gesellschaft für Technische Zusammenarbeit (GTZ) GmbH

The authors:

Burkhard Schwarz, geographer, lived mainly in Greenland, Peru and Bolivia since 1981. For the past seven years, he has been working in Bolivia as an independent consultant and researcher. At present, he is accompanying ethnic communities and an aymaran group in their self-analysis process, concerning autonomous management of natural resources and socio-cultural change.

Eberhard Gohl, economist and sociologist, spent a few years in Turkey, Peru and Bolivia. For eight years, he worked mainly with FAKT, DSE-ZEL and GTZ as a consultant for project management and organisation development. At present, he works as Controller in the German Protestant Churches' funding NGO „Bread for the World".

Dorsi Germann, sociologist and graphic artist, spent four years working in a community development project in Senegal. For the last fourteen years, she has been a consultant on adult education, appropriate technologies, technics of communication and visualization, project management, monitoring and evaluation, organizational development and participatory methods in Africa, Asia and Latin America, mainly working for GTZ and FAKT.

Die Deutsche Bibliothek – CIP-Einheitsaufnahme

Participatory impact monitoring : a publication of Deutsches Zentrum für Entwicklungstechnologien – GATE, a division of the Deutsche Gesellschaft für Technische Zusammenarbeit (GTZ) GmbH. – Braunschweig ; Wiesbaden : Vieweg.
  ISBN 3-528-02086-5
NE: Deutsches Zentrum für Entwicklungstechnologien <Eschborn>

Booklet 3. Application examples / Burkhard Schwarz ... – 1996
NE: Schwarz, Burkhard

The author's opinion does not necessarily represent the view of the publisher.

All rights reserved
© Deutsche Gesellschaft für Technische Zusammenarbeit (GTZ) GmbH, Eschborn 1996

Published by Friedr. Vieweg & Sohn Verlagsgesellschaft mbH, Braunschweig/Wiesbaden

Vieweg is a subsidiary company of the Bertelsmann Professional Information.

Printed in Germany by Lengericher Handelsdruckerei, Lengerich

ISBN 3-528-02086-5

CONTENTS

Palmyrah Tapper's Families in Tamil Nadu, India 3
1. Background 3
2. Monitoring practice prior to introduction of PIM 4
3. Introduction of the PIM concept 5
4. Details of group-based impact monitoring 7
5. Details of NGO-based impact monitoring 8
6. Impacts observed and induced by PIM 9
7. Conclusions 10

FEDECOMIN in Caracoles, Bolivia
Popular Consumption Store in Caracoles Mining Sector, Case 1: 11
1. Background 11
2. Monitoring practice prior to introduction of PIM 13
3. Introduction of the PIM concept 14
4. Details of group-based impact monitoring 15
5. Details of NGO-based impact monitoring 18
6. Impacts observed and induced by PIM 21
7. Conclusions 23

FEDECOMIN in Kantuta / Bolivia
Ore Dressing Project in Kantuta Mining Cooperative, Case 2: 25
1. Background 25
2. Monitoring practice prior to introduction of PIM 26
3. Introduction of the PIM concept 26
4. Details of group-based impact monitoring 26
5. Details of NGO-based impact monitoring 30
6. Impacts observed and induced by PIM 31
7. Conclusions 32

SIBAT / PARTNERS in Bacgong / Philippines
How PIM Came to Bacgong: An adventure in action research 33
1. Background 33
2. Monitoring practice prior to introduction of PIM 34
3. Introduction of the PIM concept 35
4. Details of group-based impact monitoring 41
5. Details of NGO-based impact monitoring 44
6. Impacts observed and induced by PIM 45
7. Conclusions 46

INDES in Misiones / Argentina
Promoting the development of a grassroots organization with PIM 49
1. Background 49
2. Monitoring practice prior to introduction of PIM 50
3. Introduction of the PIM concept 51
4. Details of group-based impact monitoring 54
5. Details of NGO-based impact monitoring 54
6. Impacts observed and induced by PIM 55
7. Conclusions 56

SUMMARY
What have we learned from the field studies? 57
1. Background 57
2. Monitoring practice prior to introduction of PIM 58
3. Introduction of the PIM concept 59
4. Details of group-based impact monitoring 62
5. Details of NGO-based impact monitoring 64
6. Impacts observed and induced by PIM 65
7. Recommendations for avoiding basic risks 66
8. Conclusions 67

Palmyrah Tapper's Families in Tamil Nadu, India
A Case Study

Amirtharat Anandhy, PWDS

The case study of Palmyrah Workers' Development Society (PWDS) is presented in order to illustrate a situation in which PIM was introduced independently and without external advisors. The NGO had only the preliminary version of the PIM guide. PIM was applied systematically, which made it easy for the communities and the NGO's field workers to understand how it works.

1. Background

Development organization / NGO

Palmyrah Workers' Development Society (PWDS) was founded in 1976. It was registered in 1977 under the Societies Registration Act. Since its inception, the Society's main objective has been to promote the formation of a movement of palmyrah tappers and other socially and economically deprived groups for their social and economic liberation and to increase their awareness, dignity and self-reliance. This is done by guiding and encouraging them in establishing community-based income-generating units, by organizing training programmes, and by participatory development of value-added palm products such as candy (sweets), syrup, spiced and refined jaggery (see glossary below). PWDS staff members come from the same social background as the palmyrah tapper families.

Self-help groups

A precondition of the development process is empowerment of the weaker sections of the population at grass-roots level. The first step towards achieving this is organization of the community. With the encouragement and guidance of PWDS, *Mantram* or village-level associations for tappers and women have therefore been formed in rural areas. The groups are involved in attaining self-sustained development through awareness education, community- based employment and income-generating programmes, and other self-help activities, e.g. a revolving loan and savings scheme. At present, 110 women's groups and 200 tapper's groups are functioning in different parts of Kanyakumari and Trivandrum Districts.

PIM is being introduced in the community-based candy production units. Community-based means that the units are owned and run by a group of palmyrah tapper families. The candy is produced in separate buildings. In the past, each family processed the neera (see glossary) in their own house, with the result that the women's workload was extremely high, children had to help in production, the houses became dirty, and the income generated by all this drudgery was very low.

Motivation, financial assistance in the form of concessional loans, and technical and marketing support are provided by PWDS. The whole programme relating to the candy production unit has been designed to promote participatory and self-sustained development.

GLOSSARY

palmyrah	=	palm which yields sugar sap
neera	=	sap tapped from the flower of the palm
jaggery	=	solidified sugar syrup
brix	=	measure of sugar content in the sap (°brix equivalent to % sucrose)

---

2. Monitoring practice prior to introduction of PIM

Development organization / NGO

Responsibility for implementing each project was assigned to different sections of PWDS, which operated with a coordinator, a supervisor and other staff members, including field workers. Planning, decision-making and monitoring were done by the superiors after joint discussions. In monitoring, the main emphasis was on assessment of activities, i.e. the technical and economic impacts of the projects.

Self-help group

Leaders selected by the group monitored the activities. Planning, decision-making and impact assessment were carried out jointly by the community and PWDS staff. The community's participation is assured in all phases from the start of production to the sharing of profits. The community holds meetings to take decisions on whether to set up candy production units, the location and management of the units, prices of neera and candy, employment of candy makers, profit-sharing etc. In short, the community is given the ability to monitor the units in accordance with PIM principles.

3. Introduction of the PIM concept

Development organization / NGO

PWDS introduced PIM gradually in its various programmes. In 1993, PIM was introduced in the palm product development programme. A planning session for staff members was held to decide on methods of implementing PIM in the candy-making units. At this session the steps in introduction were listed, and activities are now being carried out accordingly.

Next, the staff were trained in the use of PIM tools, including assessment of socio-cultural impacts. Monitoring is now being carried out on this basis.

Self-help group

Group-based impact monitoring was introduced in the community-based candy production units. PWDS provided motivation and guidance. First, the members of the community were made aware of the need for PIM, to ensure their full participation and to make the programme self-sustaining. Then they were motivated to introduce a system of this kind.

The community then articulated their expectations and fears concerning the project. Corresponding indicators were also identified. An observation committee was selected and trained in the use of PIM tools

The steps in the introduction of PIM may be summarized as follows:

Steps in the introduction of PIM in community-based candy production units

1. Introduction of community and field workers to principles of PIM
2. Articulation of community's expectations and fears
3. Identification of indicators
4. Identification of monitoring team
5. Training of team
6. Documentation of observations
7. Reporting of observations, analysis of impact and corrective action if necessary.

4. Details of group-based impact monitoring

With the help of PIM, the tappers' families are able to observe the changes brought about by the candy production units. Members of the self-help groups were trained as observers, e.g. they had to learn how to measure brix and pH of neera themselves.

All the candy production units and PWDS meet regularly at fortnightly or monthly meetings. The monitoring results are regularly reported and assessed. Impacts are analyzed and any corrective action needed is taken jointly by the community. In the course of time the system has become well established in the units.

Some of the community's expectations and fears and the indicators derived from these are listed below:

COMMUNITY'S EXPECTATIONS	INDICATORS
Higher profit from candy production than from jaggery production	Cost of producing candy; candy and jaggery prices; profit given to tappers
Production of high-quality candy with high yield	Candy yield and quality; brix of neera; pH of neera
Generation of alternative employment for women	No. of women doing alternative jobs
Promotion of saving habit	No. of members saving
Development of production unit (e.g. construction of their own building)	attendance at meeting; participation in the project

COMMUNITY'S FEARS	INDICATORS
Tappers may not cooperate in supplying neera	No. of tappers supplying neera; total quantity of neera supplied
Tappers may not supply quality neera	Brix of neera; pH of neera
PWDS may not pass on full profit to tappers	Profit given to tappers

5. Details of NGO-based impact monitoring

NGO-based impact monitoring is carried out by PWDS. PIM principles were discussed among the staff members and decisions were taken on how to implement NGO-based impact monitoring. The NGO's expectations and fears with regard to the project were listed, and indicators to measure them were identified.

NGO'S EXPECTATIONS	INDICATORS
Candy can be produced with high yield	Candy yield and quality
Gradually the community will take more and more responsibility for the unit, including analyzing impact, processing candy and maintaining accounts	List of responsibilities taken by group
Reduction of women's workload	Reduction in women's working hours
More people may join the project	Number of partners actively participating in the project
New groups may set up candy production units	List of groups with established candy production units
PO will undertake other needbased projects	List of new schemes initiated by group (e.g. community savings, community credit)

NGO'S FEARS	INDICATORS
Whether tappers' families will cooperate and become involved in the project	Number of tappers' families supplying neera
Whether sufficient quantity of neera will be available to the units	Average quantity of neera supplied daily

Observation is carried out regularly and reflected upon at the fortnightly meetings. Impacts are analyzed and corrective action is taken whenever difficulties or obstacles arise.

Joint reflection workshops are arranged for the PWDS staff and the producing groups. Experience gained in setting up and monitoring the units is shared by the people from three candy units. They discuss lessons learned regarding successes and failures of the programme and take corrective action to overcome the obstacles and to plan for future activities. Each group is able to learn from the others.

6. Impacts observed and induced by PIM

The impacts observed include not only expectations and fears. Especially at socio-cultural level the actors noticed other changes with regard to attitudes and behaviour. Some of these were induced or reinforced by the application of PIM itself.

Technical and economic impacts

1. Higher profits from candy production.
2. Efficiency of candy production is increased.
3. Tappers' income is doubled.
4. Alternative employment opportunities are created for women.
5. Women's workload is reduced.
6. School-age children have more time to devote to their studies.
7. Saving habit is promoted among tappers' families.

Socio-cultural impacts

1. Development of cooperative attitude within community.
2. Involvement of groups in decision-making, analyzing impacts etc.
3. Women are now able to devote more time to their children and loved ones.

4. Tappers' houses and surroundings have become cleaner.
5. Women have time to participate in important social functions.
6. People have better knowledge of testing and processing candy.
7. Spirit of enterprise has developed among members.

Moreover, PWDS and the communities were able to identify various multiplier effects:

integral multiplier effects
- new members are joining the programme
- women also participate more actively in monitoring: they organize meetings once a month to reflect on what has been observed and take decisions relating to the programme

vertical multiplier effects
- alternative employment opportunities are created for women
- promotion of the saving habit among the members

horizontal multiplier effects
- impressed by the success of the candy unit at Kamplar, two more groups have started producing candy.

7. Conclusions

The monitoring system being practised in the units has increased the autonomy of the groups in management. In the long run it is expected that the units will achieve full self-management.

The application of group-based impact monitoring has been a learning process for the community in the sense that their ability to produce candy efficiently has improved.

The monitoring system used by PWDS and the self-help groups can be managed easily by those involved. PIM makes it clear which impacts are important to the people and to the NGO, and helps to monitor them continuously. Although indicators may not meet scientific standards, they are helpful for the actors for the purpose of reflecting on successes and failures and thus steering the project.

If carried out properly, PIM can promote autonomy and contribute to the success of projects by augmenting growth-promoting factors and overcoming growth-depressing factors.

Address:
Palmyrah Workers' Development Society (PWDS)
Crystal Street
Martandam 629 165
Tamil Nadu
India

Lessons on Responsibility for Credit

PIM experience in mining cooperatives in Bolivia

Beatriz Delgado, Florinda Gonzáles, Iván Velásquez, Burkhard Schwarz

The case study from the **Federation of Mining Cooperatives of the Department of La Paz (FEDECOMIN)** demonstrates how people and NGO staff approach the task of clarifying the economic backgrounds and the internal organization of their joint projects. This induces individual capacity- and responsibility-building and organizational learning processes. The relevance of the work of the NGO's staff, previously underestimated, is made clear. PIM has been tested in two self-help groups affiliated to FEDECOMIN. (A detailed case study on PIM in Caracoles is available from CODECOMIN and FAKT.)

FEDECOMIN in Caracoles, Bolivia
Popular Consumption Store in Caracoles Mining Sector
Case 1:

1. Background

Development organization / NGO: FEDECOMIN with CODECOMIN

The Federation of Mining Cooperatives of the Department of La Paz (FEDECOMIN-La Paz) was founded in 1977. About 80 mining cooperatives are legally affiliated to FEDECOMIN, representing more than 7,000 cooperative members, members of their families not included.

For several years there has been a technical consultancy service within FEDECOMIN called CODECOMIN. This was developed to meet the various social and technical needs expressed by the cooperatives. Another objective of CODECOMIN is to strengthen the mining cooperatives as people's organizations.

Self-help groups: The mining cooperatives

The mining cooperatives are a genuine miners' self-help organization. The "traditional" prospect mines were abandoned by private or government companies because they were not profitable. The mining cooperatives are now a highly relevant sector of the economy, providing employment and generating national income in foreign currency. They accomplish this with very little capital and usually under extremely adverse working conditions. The average family income of cooperative miners is currently Bs. 200 - 300/month (US$ 50 - 75).

The length of the working day varies from one cooperative to another and even from one member to another. Generally they do not have sufficient breaks, and work from Monday to Sunday. The cooperatives use substandard equipment under poor conditions.

Since production is irregular, 80% of the cooperatives can save no money for internal investments. Because of their precarious economic situation all the cooperatives suffer from strong fluctuations in membership and are therefore structurally weak.

Popular consumption store in Caracoles

There are three neighbouring mining cooperatives in the Caracoles area: *Libertad, Porvenir* and *El Nevado*. In each of them there is a Housewives' Committee. While these women's organizations are formally independent, in practice they depend on the cooperatives, which are run by men. Most of the women, whose native language is Aymara, speak no Spanish.

The Popular Consumption Store was founded in 1990 as a response to the problems that cooperative members' families faced in purchasing basic necessities – mainly with regard to price and quality. The distance separating the cooperatives from the markets supplying them, the bad roads and the lack of an own vehicle were other reasons.

The founding of the Popular Consumption Store was also intended to consolidate the organization of women in the mining cooperatives in both of their roles, as active members and as housewives.

In the administration of the store, however, there were fundamental problems which made it difficult to attain these objectives. On one hand, the housewives' lack of education was an obstacle to collective and effective management, while on the other the unfavourable power structure led to continuous interference by the (male) cooperative members, e.g. regarding basic decisions on the repayment of credits.

2. Monitoring practice prior to introduction of PIM

Development Organization / NGO: CODECOMIN

CODECOMIN mainly accompanies productive and social projects. While the work done has been monitored regularly, comparing it to the plans was not very systematic and was hampered by frequent problems and fluctuations in the cooperatives and by changes in the general political situation in Bolivia.

The monitoring mechanisms were not sufficient to reflect changes and impacts within the projects. The information was restricted to a small group of individuals, i.e. staff members and federation leaders. Three-quarters of the cooperative members may have noted some changes individually, but there was no opportunity to analyze them collectively: were they positive or negative, were they due to members, leaders or staff? There was a lack of communication between cooperative members and staff.

CODECOMIN was unable to understand the logic of internal functioning and the complexity of the cooperatives' organizational problems. Due to this unstable context, practically none of CODECOMIN's projects evolved as planned. This situation caused frustration among the staff members, the cooperatives, within the federation, and even with the funding agency. The work of the staff, and especially of the social development department, was not at all appreciated.

Against this background the socio-economic consultant of FAKT suggested that CODECOMIN should participate in the PIM field phase. PIM was supposed to improve project guidance by making monitoring more participatory, more realistic, more successful, and by making the hidden achievements of the CODECOMIN team visible and thus more respected.

Self-help group: Housewives' Committees in Caracoles

The Caracoles Housewives' Committees undertook general monitoring, but not with regard to the store, the purpose of which was not only to provide goods but also to improve nutrition in the miners' families. Formally, the women's store was managed through regular meetings of the housewives' committees on the basis of quarterly financial reports. In practice, the meetings tended to be spontaneous and restricted, involving only a small group of housewives. Management was not continuous. Nor was it transparent, as it was concentrated in the hands of a few very dynamic individuals. The men from the cooperatives intervened by fixing the prices, while administrative tasks were delegated to university students doing practical training in the mines.

The administrative reality was far from the aspirations expressed in the objective to make the housewives active managers of the store and thus to stimulate and strengthen the Housewives' Committees.

3. Introduction of the PIM concept

Development Organization / NGO:
Development of a preliminary PIM concept

As a first step in implementing PIM at FEDECOMIN/CODECOMIN level, a workshop was held with federation leaders, one representative each from Kantuta and Caracoles, and staff members from CODECOMIN's Social Development Department. At this meeting it was finally decided to implement PIM in the two cases reported on here. The new spirit of socio-cultural participation to be introduced by PIM was articulated by starting the workshop with two Andean ritual ceremonies, the *akulliku* and the *ch'alla*.

Afterwards the participants exchanged ideas about the results and limitations of planning, and about conventional monitoring techniques.

Then they asked themselves:

- What is PIM good for?
- What do we want to achieve by implementing PIM?

To obtain answers to these questions, they held a brain-storming session and recorded the ideas.

Self-help group

In February 1993, several formal and informal meetings were held to introduce PIM in the context of the Popular Consumption Store - with the housewives, the members of the cooperative and both genders. Those attending the meetings set up an appropriate indicator system.

Also in this context, the *akulliku* and the *ch'alla* were indispensable for consolidating communication and creating the right atmosphere. In addition, various popular education methods were applied to avoid weariness. It helped the participants to express their opinions more freely. At first it was mainly the men who voiced their opinions concerning the problems, but later more of the housewives participated in several spontaneous brainstorming sessions, putting their views on the Popular Consumption Store.

4. Details of group-based impact monitoring

The indicator system established by the **Libertad, Porvenir** and **El Nevado** cooperatives was defined in three steps. A formal meeting was organised for expressing project expectations and fears. Their points of view were recorded on poster paper. For the purposes of this booklet we have selected just a few of the 14 expectations and 9 fears recorded:

EXPECTATIONS	FEARS / DOUBTS
that the nutritional status of cooperative members would be improved	that members of the cooperative would not pay their debts
that there would be active participation of housewives	that it would not be supervised by the housewives' committee
that basic foods etc. would be supplied at fair prices	that they would not be able to administer the store and would go bankrupt
that the housewives would learn how to run a store	that they had a total lack of knowledge of how to run the store

Identifying indicators

In a second step, through discussion, the participants reduced their expectations and fears to observable indicators, again using poster paper:

EXPECTATIONS / FEARS	INDICATORS (derived from expectations or fears)
that the nutritional status of cooperative members would be improved	Increase in consumption of particularly nutritious food products
that basic foods etc. would be supplied at fair prices	Prices are fixed according to increase in market prices
that there would be active participation of housewives that it would not be supervised by the housewives' committee	Housewives' Committees are informed about store management at each meeting
that members of the cooperative would not pay their debts	Names of cooperative members who have not paid and the amount they owe are recorded

In total, 14 indicators were identified, and of these 8 were prioritized. The other indicators should be kept in mind for further studies.

Determining the method of recording indicators

The third step resulted in the following recording method for each indicator (only an extract can be documented in this booklet):

INDICATORS (derived from expectations or fears)	OBSERVATION METHODS
Increase in consumption of particularly nutritious food products: – higher consumption of lentils and quinua – lower consumption of noodles, rice and sugar	Measure weight of each product sold (lentils, quinua, corn flour, dried field beans, noodles, rice and sugar) separately, and record it appropriately
Prices fixed according to increase in market prices, with coordination of the three cooperatives	Answer YES or NO and record observations and comments
Housewives' Committees informed about store management at each meeting	Answer YES or NO and record observations and comments
Names of cooperative members who have not paid and the amount they owe are recorded	Keep a record in the form of a table, separate for each cooperative

Some indicators were recorded in tables, others also as graphs. One reason why we preferred to visualize some figures was that changes would be understood better by housewives who were not used to reading.

Building an observation team

Next day, the meeting was continued in order to choose the observation team, which consisted of three housewives, one from each Housewives' Committee or cooperative.

Final definition of observation instrument

Simultaneously, a meeting was held with the observation team of the three cooperatives, to train the team in using the observation instruments, e.g. how to draw a graph. At this meeting we defined in detail the method of recording the indicators prioritized at the previous formal meeting.

Informal meeting: women speak out

Afterwards a meeting was held, chaired by a cooperative leader. Each cooperative put its point of view on the management problems of the Popular Consumption Store and possible solutions to them. The great majority of the participants were women. The problems and solutions which they discussed were simultaneously visualized by the moderating leader. The results of this discussion helped to define the indicators of the group-based PIM.

5. Details of NGO-based impact monitoring

The indicators were derived from the expectations and fears of CODECOMIN members concerning the running of the Popular Consumption Store.

First, the participants expressed their expectations and fears with regard to the Popular Consumption Store in a verbal brain-storming session, which was recorded by the moderator on poster paper. As a second step, after a short discussion, the participants reduced these expectations and fears to observable indicators, again visualized on poster paper. The indicators listed were later prioritized; the other indicators were to be kept in mind for further studies.

Next the participants agreed on methods of recording observations. They developed tables for the measurable indicators, and for the descriptive indicators they set up YES/NO-questions, asking for additional "observations". The watchers were elected, and the equipment they needed was ascertained.

However, as a result of the dynamics of the meetings in Caracoles, a series of modifications were necessary. Some days later, the CODECOMIN-PIM team held a team reflection session on the activities accomplished in the introduction of PIM in Caracoles. As numerous interesting aspects of management of the store had come to the surface at the informal meeting, some of the conclusions to be drawn were now discussed.

The following is a selection of the team's expectations and fears:

EXPECTATIONS	FEARS / DOUBTS
that prices would be lowered and better quality offered than by other stores	that the cooperatives might not pay the store
that more integration of the three Caracoles cooperatives would be achieved	that the women might not be able to run the store themselves
that the Housewives' Committees would take over responsibilities	Should they sell for cash only or give credit?
that spaces for the participation of women would be created in the cooperatives	that they would not recover credit

EXPECTATIONS / FEARS	INDICATORS
that prices would be lowered and better quality offered than by other stores	The prices of 20 staples in the store are lower than in other stores nearby
that they would not recover credit	The total credit given each month does not exceed the cash payments received
that the cooperatives might not pay the store	
that more integration of the three Caracoles cooperatives would be achieved	A meeting of the cooperative leaders to be held each month to analyze the recorded results
that the Housewives' Committees would take over responsibilities	The committees have been able to efficiently control the discount granted

The CODECOMIN staff selected 11 indicators for observation, of which 9 were prioritized. The following four have been chosen as examples:

INDICATORS	OBSERVATION METHODS
The prices of 20 staples in the store are lower than in other stores nearby	List the prices of each staple monthly for the store and for 5 other stores nearby separately; Answer YES or NO and record observations and comments
The total credit given each month does not exceed the cash payments received	List the credit granted and payments received. Answer YES or NO and record observations and comments
A meeting of the cooperative leaders to be held each month to analyze the recorded results	Answer YES or NO and record observations and comments
The committees have been able to efficiently control the discount granted	Price lists and prices charged are checked monthly by the committees. Answer YES or NO and record observations and comments

PIM 3 · Application Examples

Possible further indicators

The informal discussion with the women in Caracoles on the store's problems and possible solutions to them produced plenty of information. From this information the team derived possible indicators and agreed to keep them in mind (the following is an extract):

INDICATORS	RECORDING METHOD
The established business hours have been maintained	Answering: YES - NO Observations
Opinions of informal groups have been recorded	Answering: YES - NO Observations
Interventions of management in opposition to agreements concerning credit granted by the Cooperative's Store were avoided	Answering: YES - NO Observations

Joint reflection meetings

From the very beginning, joint reflection meetings were intended as the central element in the application of PIM. They were therefore to be held monthly, because it seemed necessary to support the observation team methodologically and also to strengthen their position within the local power structure, which in all three cooperatives was opposed to the store being run independently by the housewives.

With regard to the performance of PIM, an internal evaluation meeting was held at CODECOMIN level to analyze it at both levels. It was decided to continue with the same indicators. New indicators were developed from the accumulated information about the store. These potential indicators were to be taken up as concrete proposals at both PIM levels if it proved necessary to complement PIM and complete the observation - reflection - action cycle more efficiently.

6. Impacts observed and induced by PIM

Many things have changed since the introduction of PIM. Impacts of the Popular Consumption Store project have been observed and in some cases even directly induced with the help of PIM.

Impacts on the project

Among the housewives and the miners of Caracoles there was increasing concern about the devolution of credits. Not only did the miners repay their credit to the store; the store also increased its amortization of the credit which had been channelled by FEDECOMIN/CODECOMIN. The housewives took more care to adjust the prices in the store regularly. In time the store received new supplies more regularly and its stocks of nutritive staple food increased.

Although many of the housewives were illiterate, their organizational structures became more dynamic. Women participated more actively in the project and in solving the problems of their store, and this led to a higher estimation of their own abilities and thus to greater self-esteem.

Their real ability to solve problems also increased; new, positive attitudes towards the store emerged, and the women started to take their own medium-term and long-term aims into account. This was manifested when they introduced new administration rules which were more transparent and appropriate for the cooperative structures; this again increased the demand for more transparency in the work of the housewives' committees.

The importance of communication regarding the joint project within and between the organizations increased both in CODECOMIN and in the Housewives' Committees. All this led to a new awareness on the part of the housewives about their own capabilities, and as a result the economic situation of the store gradually improved.

Impacts with regard to PIM

PIM generated mechanisms to feed back information about the store between the genders and to facilitate dialogue between leaders and members. PIM helped make all these changes more visible and contributed to collective reflection on them (among both housewives and miners).

The Housewives' Committees do not yet monitor independently, support from CODECOMIN is still needed. But PIM has put more social pressure on the leaders to penalize lack of responsibility with regard to the store, and it has dynamized decision-making and adjustments in the store.

Many cooperative members did not like PIM because of its tendency to show up existing problems and provoke confrontation. The housewives, however, recognized that PIM helped to analyze and resolve the problems of the Popular Consumption Store. PIM seems to work as a continuous learning and reflection process and is thus becoming a catalyst for the sustainability of change.

7. Conclusions

It is important to point out that the way an indicator is perceived depends on the context in which it is used.

The first indicator of group-based PIM (*Increase in consumption of particularly nutritious food products: – higher consumption of lentils and quinua, – lower consumption of noodles, rice and sugar*) aims at resolving a sociocultural problem, as it focuses on new consumption patterns to improve the nutrition of the families in the mining cooperatives.

Before the PIM application, a certain apathy of the housewives was noticeable in that they did not ask for such products either in the Popular Consumption Store or in other shops. After the introduction of PIM they started to ask for even more products with high nutritive value, not only in their own store but also in the other shops. There was an appreciable increase in the consumption of quinua, lentils and corn flour among the families in Caracoles.

From the fourth indicator of group-based PIM (*Names of cooperative members who have not paid and the amount they owe are recorded*) may seem to be a technical and economic indicator; from the housewives' point of view, however, it reflects the power structure.

This indicator was questioned from the start, mainly by the (male) cooperative members, and it was even a topic of open disputes between the housewives and the miners. It is of central importance because it visualizes, on a large chart in the store, how much each member of the cooperative owes the store. While the women were at first afraid to display this sensitive data in public, after some months the observation team, with the support of all the committees, deliberately published the lists inside the store: the fact that every member of the cooperative could see who owed how much whenever he came to the store led to a certain collective consciousness and increased social pressure, with the result that the miners repaid their debts faster. This transparent monitoring even caused problems between husbands and wives, but fortunately these were resolved in due course.

Something similar happened with other indicators from group-based PIM. In this case it was observed that the cooperatives continued not to allow women to attend its meetings. It was thus impossible to report on the Popular Consumption Store at the cooperatives' meetings. (Reports would have been necessary, as the cooperatives also had debts with the store, which they did not pay back.) Consequently, the women changed the indicator slightly (monitoring the presentation at the meetings of the Housewives' Committees instead of at the meetings of the cooperatives).

It may be concluded that if the people themselves select the indicators to be monitored, no clear distinction can be drawn between technical and economic indicators on the one hand and socio-cultural indicators on the other. Selecting the subjectively important changes to be watched regularly means starting to monitor the socio-cultural impacts upon oneself.

Address:
FEDECOMIN/CODECOMIN
Casilla 11394
La Paz
Bolivia

Burkhard Schwarz
Casilla 395
Oruro
Bolivia

FEDECOMIN in Kantuta / Bolivia
Ore Dressing Project in Kantuta Mining Cooperative
Case 2:

1. Background

Ore Dressing Project in Kantuta

Kantuta Ltd. is a cooperative which runs a gold mine. It consists mainly of people who were previously employed as mineworkers, drivers, bricklayers and in some cases university graduates. At first there were 50 members, but because of the working conditions and the low profit margins this number gradually decreased. Now only 26 members are left and almost all of them live with their families in the city of La Paz, a six-hour journey from Kantuta. The miners work in two shifts: one half of the members are working in the mine while the other half stay in La Paz. The change of shift once a month is the only opportunity for a meeting of the whole cooperative.

Kantuta cooperative has invested about US$ 200,000 in improving and enlarging the equipment and machinery in the mining and mineral concentration process. By increasing the exploitation rate the cooperative would be able to solve the problems of low productivity. The miners of Kantuta participated in the planning of the investment project. CODECOMIN was involved, but only its Mining Department, not its Economic or Social Development Department.

However, the new equipment is not working at full capacity and Kantuta's investment project is facing several problems. At a meeting with FEDECOMIN/CODECOMIN, the members of the cooperative therefore decided to implement Participatory Impact Monitoring (PIM).

2. Monitoring practice prior to introduction of PIM

Once a month there was a meeting of the cooperative at which the members were informed about progress in implementation of the investment project. One member was responsible for financial management. Reporting was both formal and informal but not very systematic. The cooperative's leaders relied on the technical reports from CODECOMIN's Mining Department.

3. Introduction of the PIM concept

Self-help group

At the various meetings between the CODECOMIN-PIM team and the representatives of the Kantuta cooperative, several participatory and motivating methods were applied. The akulliku and ch'alla ceremonies aroused the miners' interest and not only helped to establish a basis of confidence, but also facilitated spontaneous brainstorming sessions, allowing every participant to express his opinion about the ore-dressing project.

4. Details of group-based impact monitoring

Expressing expectations and fears

The indicators were defined in two steps. They were derived from the members' expectations and fears relative to the investment project. The members' opinions were expressed in a verbal brain-storming session and recorded on poster paper by the moderator. The following table is a selection from the total of 14 expectations and 14 fears recorded:

EXPECTATIONS	FEARS
that the equipment and machinery would be improved to increase production	divergent opinions on the convenience of the credit, depending on individual members' means
that living standards would improve (working hours, new dining room)	that the credit would not satisfy their needs
that the miners would have less hard work	that production costs might increase
that there would be confidence in the members' responsibility with regard to the fulfilment of their credit obligations	that the members would be frustrated because of the delay in granting the credit

Identification of indicators

In a second step, the participants discussed their expectations and fears and reduced them to observable indicators on poster paper. They were classified according to the categories to which they belonged (technical, economic, social and organizational areas). In some cases several expectations and/or fears were summarized in a single indicator.

A few days later a meeting was held with the observation team of the Kantuta cooperative to train them in application of the observation instruments. They defined in detail the method of recording the prioritized indicators. The result of this session was a series of forms to be filled in by the watchers.[1]

---

[1] At the (both formal and informal) meetings opinions were expressed which revealed greater and broader expectations than documented in the study.

Examples:

Fear: Divergent opinions on the convenience of the credit, depending on individual members' means.

Indicator: Recording of opinions about the cooperative's investments (at formal and informal meetings):

RECORDING OF OPINIONS

Formal meetings	Informal meetings
– we should invest in new machines to improve production	– no clear idea why production is lowering
– mill does not work well → losses	– members don't cooperate to optimize production

Observations: ....

Expectation: That there would be confidence in the members' responsibility with regard to the fulfilment of their own credit obligations

Indicator: Payment of interest on the credit was regularized by paying US$ 4,000 by the end of the month.

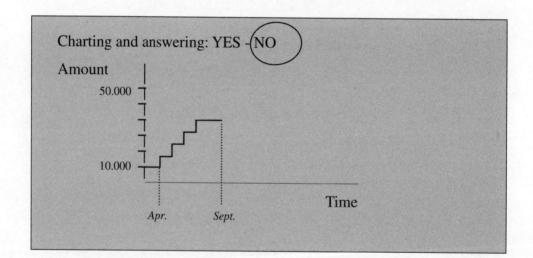

Observations: No. Due to the low production, it was only possible to pay 1,000 $ last month.

Expectation: That the equipment and machinery would be improved to increase production

Indicator: The operation of the equipment and machinery bought with project funds was gradually improved.

Recording in a table:

Equipment / Machinery	Problems identified
Grinder	does not work
Radio	works at 100 %
Transport	works at 100 %
Mill	works at 20 %
Jigs	works at 70 %
Perforators	one does not work

Observations: high fuel consumption for generator

Expectation: That the miners would have less hard work
Indicator: The work done by the members will gradually be mechanized.

Recording new acquisitions of equipment and machinery, and answering:

(YES)- NO

1 compressor was brought

Observations: investments specially inside the mine

PIM 3 · Application Examples

5. Details of NGO-based impact monitoring

After the meeting with the miners from Kantuta, the CODECOMIN-PIM team reviewed its expectations and fears relative to the investment project. For the purpose of this booklet we will document only four indicators and present examples of observations made from a monthly monitoring sheet:

1. The person responsible presented monthly reports on the project at every meeting.

   Observation: Yes, the current situation of the Investment Project has been mentioned at every meeting, but this has not led to any decisions.

2. The instalments and the membership fees of FEDECOMIN are up to date.

   Observation: No, the credit repayments have not been regularized. Yes, the payments to FEDECOMIN-LP are up to date.

3. The number of members of the cooperative has increased.

   Observation: No, because the value of shares has not yet been decided.

4. The Cooperative's book-keeping is up to date.

   Observation: Yes, since January 1993.

6. Impacts observed and induced by PIM

As in Caracoles, many things have changed since PIM was introduced:

PIM has **introduced new dynamics** into the cooperative: people are aware that it is necessary to prepare meetings beforehand. The miners are anxious to bring the indicators up to date and incorporate new ones in order to document the new situations and problems. They now want transparency in the economic management of the investment project, so that the organization structures in general are being made dynamic and updated.

The cooperative has introduced new administrative procedures which are more transparent and more appropriate. The permanent recording of changes facilitates decision-making on necessary adjustments of organizational structures. Despite an initial conflict on competences, the members are continuing with PIM because they know they have to optimize their organization structure.

In particular, the miners of Kantuta have recognized the importance of communication, both for processes within their cooperative and personally among themselves. The professional resources of the members who have graduated from universities are valued more highly. However, the members have also become aware that in the past they did not undergo sufficient training.

Generally, PIM facilitated participation in project activities. The miners gradually abandon their short-term view in favour of achieving their long-term goals. They develop attitudes and the capability to confront problems and not postpone them. They are increasingly aware how important it is for all members to participate in the investment project and a general responsibility to comply with the obligations of the credit has emerged. The reports from the joint reflection meetings are also highly valued by the funding agency (EDCS) because they give a picture of the cooperative's situation.

Also, the **attitudes towards monitoring have changed**, in two directions:

Negatively, because PIM makes problems visible and thus brings conflicts to the fore. Therefore, it is still not easy to maintain the new monitoring rules autonomously: PIM is still dependent on the motivation of CODECOMIN staff.

Positively, as people recognize that PIM helps them in analysis, criticism, self-criticism and responsible implementation of the tasks assigned to them, and they consider it the best way to overcome their problems. The members of the cooperative demand administrative continuity, avoiding untimely interruptions, because they know that they can only solve their problems and turn in a good performance if their leaders achieve their tasks.

The joint reflection meetings give prominence to the quality of the individual and thus motivate the members to participate. The CODECOMIN staff attaches more importance than in the past to the moments of informal communication. At the meetings, the quality and dynamics of decision-making has improved, and an awareness of "time investment" and efficiency has developed.

7. Conclusions

The FEDECOMIN leaders' and the CODECOMIN staffs' experience with PIM in Caracoles and Kantuta motivated them to introduce an improved, systematic monitoring system for all CODECOMIN programmes. This new monitoring system is a mixture of conventional monitoring, derived from planned objectives, and NGO-based PIM, derived from the staff's expectations regarding socio-cultural changes.

Address:
FEDECOMIN/CODECOMIN
Casilla 11394
La Paz
Bolivia

Burkhard Schwarz
Casilla 395
Oruro
Bolivia

SIBAT / PARTNERS in Bacgong / Philippines

How PIM Came to Bacgong: An Adventure in Action Research

Resource Persons:
Amando A. Bolunia, Dorsi Germann

1. Background

The two development organizations / NGO

SIBAT is a network of NGOs based in Manila, Philippines. It gives advice and support to its member NGOs in the fields of Appropriate Technologies and Sustainable Agriculture. The services offered are an information and library service, project development assistance, and technology assistance. SIBAT's principles are sharing experiences and responding to needs. For many years SIBAT has been a cooperation partner of GATE/GTZ.

PARTNERS Inc. is an NGO and a member of SIBAT. Its working area is the province of Camarines Sur. PARTNERS was willing to participate in the PIM field phase and selected the community of Bacgong, where a project called the Agricultural Production Project (APP) is being implemented.

APP has two major components:

1. Organization and education.

2. Agricultural production (rice, vegetable and livestock) to improve nurition and self-sufficiency.

Self-help group (SKB)

The community of Bacgong has 67 households with a total of approximately 360 individuals. The main economic activity is agriculture. Food production is below subsistence level. The community consists of two types of organizations: the *barangay* as the basic unit of government in the Philippines; and SAMAHANG PANG-KAUNLARAN NG BACGONG (SKB), which was organized as a people's organization or a self-help group of farmers in 1989 by staff of PARTNERS.

SKB's main task is to carry out projects. The organization has three sections the Farmer, Women-farmer and Youth Sectors. Practically all the inhabitants of Bacgong are members of SKB. For the purposes of implementing PIM, the Women's Production Team was chosen to work on the project because of its group cohesion, manageability, and level of articulation of every member of the group.

There are thus three organizations involved in this PIM project:
1. SIBAT Network Secretariat
2. PARTNERS Inc.
3. SAMAHANG PANG-KAUNLARAN NG BACGONG (SKB): a self-help group

2. Monitoring practice prior to the introduction of PIM

Development organization / NGO

In theory, monitoring consisted of verifying planned activities, their status and any deviation from the plan. In addition, new opportunities were to be identified on the basis of project experience. In practice, however, there was more talking about and planning of monitoring than actually doing it.

Reporting is based on correspondence, narrative reports and tabular reports. There are quarterly and annual reports. The difficulties in report writing lie in constructing data („What is there to write?") and in explaining the status of the activities and the projects. Above all, the complexity of a project is difficult to describe.

The monitoring system is time-consuming, the data are unreliable, and the system does not allow timely intervention in the projects.

Self-help group (SKB)

Monthly meetings are held to discuss the problems in the organization; the analysis of problems is unsystematic. Monitoring is mainly done in informal discussions, by sharing stories.

Every three months, the situation is described and assessed on two to three pages. This is summarized in a form with four columns:

PROJECT	WHAT HAS HAPPENED?	PROBLEMS ENCOUNTERED	RECOMMENDA- TIONS
cucumber 1/4 kg seeds	seeds were distributed to all 3 groups	leaves of some plants turned yellow	in future a full-time technician should check the plants for disease
....	....	....	....

The area coordinator of PARTNERS collects these sheets. However, until now they have not been incorporated in an efficient feedback system.

3. Introduction of the PIM concept

Development organization / NGO

The fact that three different organizations were locally involved in the implementation of PIM meant that the process took longer. But it had the advantage that in SIBAT and PARTNERS there were two equal organizations which depended little on one another.

Only SIBAT had been a cooperation partner of GATE for many years, and was therefore committed to the PIM project. PARTNERS was independent and only associated with SIBAT via the network.

As the Network Secretariat, SIBAT rarely implements projects of its own, it mainly cooperates with member organizations. SIBAT therefore needed the consent and cooperation of PARTNERS to implement and test PIM.

For this reason PARTNERS had to be convinced right at the start that PIM was a feasible and useful method; so the basic concepts and ongoing strategies of PIM had to be well understood and accepted.

PIM therefore required an easily understandable and smooth means of information transfer for different levels (NGO and self-help group) which would avoid generating resistance or apportioning blame (for concerning gaps in personal knowledge or mistakes in project implementation). Moreover, it should not be inconvenient, boring or lengthy: PIM had to be as attractive as possible.

It was therefore decided to use visualization. Many parts of the PIM concept were translated into drawings and picture-stories related to everyday life in the Philippines. This was done together with the colleagues of the SIBAT Secretariat and PARTNERS, to some extent directly at village level. So by preparing the various PIM workshops, the highly abstract information was transformed into concrete cases. Participants learned by translating verbal into visual messages. A teacher-pupil situation between colleagues was therefore avoided. The situation was less tense and work and pleasure were combined.

Two farmers growing corn

36 PIM 3 · Application Examples

The pictures and stories helped to situate PIM in a familiar local context. The visualized message stimulated learning processes through both cognitive and affective responses. Humour and fun helped to reduce anxiety caused by too much pressure to understand the theoretical concept. Boredom and resistance in the face of routine tasks were also reduced to some extent.

PIM was first introduced to the staff members of the SIBAT Secretariat who were involved. After levelling-off, the second step was carried out by preparing the workshops to introduce PIM to PARTNERS and to the community of Bacgong.

To introduce PIM to PARTNERS, a two-day meeting was held with the staff involved. The intention was not merely to give an idea of what form PIM should take at the project site, but also to agree among partners about the strategy to adopt in carrying out this activity together with the self-help group (SKB).

PROGRAMME FOR MEETING WITH PARTNERS

OBJECTIVES:

1. Level off with PARTNERS and SKB member on what PIM is
2. Come up with a mechanism for operationalizing PIM

PROGRAMME:

1. Introduction to PIM

Purpose of visit / Context of PIM / Actors in PIM

2. Checking expectations / Setting objectives

3. Presentation of PARTNERS' and SKB's experiences in establishing and running a project

- Aim: to acquaint SIBAT with the dynamics of PARTNERS' and SKB's projects

- Entire project cycle; how does PARTNERS start projects?

- Guiding questions:
    1. What important changes for the people has your work induced?
    2. Which changes are normally reported on? Which of them are often not evaluated?
    3. What has changed in people's behaviour? What have they learned?
    4. Have other groups learned from this experience?
    5. Is it possible to find simple indicators for these changes?
    6. How far can these indicators be observed by the group members?

4. Presentation of Participatory Impact Monitoring

- participation, monitoring, impact
- introduction of PIM
- project timetable

5. Setting up a mechanism for implementing PIM

- the workshop intends to come up with a set of working objectives for the project to work on
- in discussing the topic, the workshop should consider that the actual working mechanism of the project will be determined and designed by SKB. What the workshop could discuss are the initial steps that will help SKB bring PIM up to date.

This programme was not adhered to in every detail. As APP is the first joint activity of PARTNERS and SKB, many questions were still open. There was a general agreement to start PIM, but SKB's capability to carry out monitoring of this type was questioned, in view of the fact that they have little experience so far in development projects.

A number of experiences regarding participatory development surfaced during the discussion:

1. Appraising socio-cultural changes in a community requires a people's organization with fairly well developed awareness and lengthy experience of community projects. Otherwise there is a danger that observation/data gathering might be fixed on the technicalities of the project.

2. The PIM project presents an opportunity for an NGO to systematize and document and gather data on the effects of its services. On a large scale, NGO monitoring of the people's organization is more a „black box", depending on individual subjective perception rather than on a specific set of criteria applied to everyday reality of the group.

3. There are technical and behavioural constraints to people's participation in development work, above all to participatory management: e.g. shortage of skills, absence of a critical awareness necessary to deduce and induce cause-effect relationships, and an absence of interest and motivation manifested by an attitude of helplessness and inability to influence a given situation. This must be resolved before effective project management and implementation are possible. There are a few individuals in the people's organization who can do the job, but training involves years of experience before any tangible results are visible.

4. The presentation of concepts should be simple, easy to understand and adapted to the socio-cultural context.

Self-help group

A first workshop was held for two villages. PIM was explained in simple form by drawings, picture stories and referring to other examples of day-to-day life. Supplementary visualized materials were distributed. The text was written in the local language. The main objective was to raise awareness and attract one community for cooperation. SKB from Bacgong agreed to participate in the PIM-project.

A second workshop was then held in Bacgong for SKB only. The participants were asked about their expectations and fears; simple indicators were then identified to observe some of these fears and expectations.

Setup of project expectations and fears

EXPECTATIONS	FEARS
that the project will promote the development of the community	that crops will not grow properly
	that labour and other efforts will be wasted
that the project will help to improve the livelihood of the people in the community	that the crop yield might not be bought at the right price
that we will no longer need to buy vegetables from the market	that it will be difficult to deliver produce to the market because there is no reliable bus service
that other barangays will not have to buy vegetables from the market	that the military might confiscate the crop
that other barangays will follow in having an organization involved in production	that we might not be able to pay back our loans for the crop
	that pests might get on the crops
that cooperation in the organization will be better and smoother	that the harvest will be poor because of drought
that crops will be improved	that the harvest might not be sold

Two aspects were chosen for observation: the social development of the group and its agricultural production. Between these, 2-3 expectations/objectives were selected by prioritizing. At this very early stage only a limited number of topics were suggested for monitoring, to avoid overburdening the farmers. It was important not to diminish the motivation to start something new. Other aspects could be added later when people were more experienced.

4. Details of group-based impact monitoring

In accordance with their own expectations and fears concerning the APP project, the SKB members decided to monitor the two aspects with several indicators:

1. Improved cooperation

INDICATOR	SOURCE	FREQUENCY
percentage of group attendance at SKB's meetings and activities	verification of attendance reports	once a month
compliance with organizational schedule	verification of organizational plan of activities	once a month
efficiency of activities	verification of activity reports	once a month

2. Agricultural production and consumption
("Food Always In The House" - FAITH-programme)

INDICATOR	SOURCE	FREQUENCY
source of vegetables for domestic consumption	locality / source of vegetable	weekly
varieties of preparing particular vegetable dishes	names of new dishes / vegetable preparation	weekly

Two members of SKB were nominated to observe and regularly document these indicators. During the second visit, however, it became clear that observation and documentation had been rather irregular.

Expectations and fears were then reviewed, and the discussion revealed that the increased vegetable production had led to a surplus which could not all be consumed in the village. The farmers expressed the wish for direct marketing and asked for assistance from PARTNERS. The NGO, however, refused to support marketing as the FAITH programme's objectives were improved nutrition and self consumption – not marketing.

The project reality had changed. Some of the objectives of the FAITH project were no longer compatible with the farmers' expectations. Nevertheless, the NGO and the farmers' group agreed to continue PIM under these conditions. The reinstallation of a project-oriented, non-adapted monitoring scheme forced the farmers to choose items in which they had little or no interest.

1. Smooth and improved cooperation in the organization

Month: ... Indicator	No.	Problems	Recommendations
monthly attendance			

2. Not to buy vegetables from other places/markets

Month: ... Vegetable	Individual Production (quantity)	Consumption (quantity)	Source

This second attempt also failed. There was no need to observe, reflect and act on items that were unrelated to the farmers' needs. There was no motivation. The hope with which the farmers had begun their cooperation with the NGO gave way to disillusionment.

In the meantime, SIBAT and SKB carried out a Participatory Rural Appraisal (PRA). In the light of the problems identified and strategies worked out together, changes in the APP project appeared advisable, too. PARTNERS felt that they had been passed over.

PARTNERS, which bore most of the responsibility for implementation of the APP project, finally saw the need to change the concept of the project. They decided to start a new programme for Bacgong, introducing a demonstration farm. However, neither the areas chosen nor the changes in the project concept were suited to actual needs. As a result the monitoring again made no progress and failed.

As there was no successful ongoing project, the community of Bacgong decided to concentrate increasingly on internal processes, i.e. the gradual increase in the cohesion of the group (by counting the participants at meetings) and the joint activities to resolve problems at village level.

One fear was: „The crops in our gardens are often eaten by stray animals". This caused in-depth discussions in the group. The Barangay Council had passed an ordinance banning loose livestock, and this had worked for a certain time, but then some owners released their animals again. At the meeting, the group described the community before and after the ordinance was enacted:

BEFORE	AFTER
1. The community pavement was littered with animal dung.	1. The pavement is unlittered.
2. There are no gardens outside houses.	2. Gardens can be found in household yards.
3. Farm plots were located far from the community.	3. Farm plots are nearer than before.

The number of households with stray pigs was counted. It revealed that there were only three of them, two of which were already looking for their pigs that night. The problem had been resolved instantly by the group itself.

The self-help group of Bacgong continuously experienced that many of their expectations were not fulfilled by the APP project and by PARTNERS. Other personal demands and requests did not fit into the framework of the project. By observing/monitoring their own activities and by reflecting that they were often solving problems themselves, their belief in the omnipotence of the NGO slowly diminished, and criticism and self-confidence grew.

The farmers became more active. They decided not merely to observe group cohesion by attendance of participants but by observing how the meeting was held.

Typical questions were:

- Who talked and how much?
- Who dominated the meeting?

They developed a documentation system of their own: to visualize the length of a speech and the dominance of one speaker they drew increasing sizes of a mouth behind his name.

Increasing importance was given to the aim of learning and growing until they themselves would be able to cooperate directly with a funding agency: they started discussing how they could learn and grow and how these procedures could be observed and documented.

5. Details of NGO-based impact monitoring

The NGO's existing monitoring system was initially regarded as adequate, although it was not very efficient in the case of the APP project. Even the visits to the community were quite rare, mainly due to the distance and the time it took to walk there.

The expansion of the existing monitoring system was rediscussed when the PO changed its own monitoring practice and decided to watch the cooperation pattern instead of technical issues only. It was agreed that there was a need to know about, and reflect on learning processes at grassroots level. The questions arose as to whether and how a self-help group could initiate and observe its own learning processes and how these could be accompanied and monitored by an NGO within an acceptable time-frame.

The NGO agreed to examine these questions, to clarify details of an additional monitoring scheme and to inform the SIBAT Secretariat. But in the end a broader, NGO-based impact monitoring system was not established.

The argument was that it was not considered necessary to monitor similar aspects as monitored by the people's organization. The need for knowledge and discussion of the learning processes at grassroots level was agreed upon, but the time needed was considered to be too high.

Other possible reasons why NGO-based PIM was finally rejected include
- the unsolved problems between SBK and PARTNERS
- the open conflict between SIBAT and PARTNERS
- the underlying doubts about the usefulness of the APP project
- the inability to accept and recognize weak points and mistakes and to learn from them
- the fear that a transparent, participatory monitoring scheme would make it easier to control the group from the outside.

6. Impacts observed and induced by PIM

At the level of the self-help group SKB the SIBAT staff observed the following impacts:

1. even a weak project can mobilize and initiate capacity-building among farmers if sound methods of analysis and reflection are applied
2. skills were developed in using and managing tools for visualization
3. awareness of the management of the organization has increased
4. self-esteem and self-assertiveness increased
5. the farmers have started to develop their own tools
6. the farmers have moved from resignation to increasing intervention in their own affairs
7. the farmers have recognized that learning and growing are very important
8. the farmers wanted to become equal and independent partners of the NGO and the funding agency (FA).

At NGO level, the impacts noticed were:

1. SIBAT became aware of the project performance and the inadequacy of the existing management system
2. documentation was recognized as refined information for decision-making
3. SIBAT and PARTNERS became aware that awareness-raising and the development of skills need not be long-drawn-out processes:

tangible results can be seen in small situations in the community
4. it was shown that ineffectiveness is more likely to be due to the method of intervention than to a lack of intervention (quality instead of quantity)
5. PARTNERS became aware of the discrepancy between the objectives of the APP project and the expectations of the farmers' group
6. through PIM, all the actors involved gained further insights into the APP project, including its weaknesses; but the fact that positions were attacked and defended caused conflicts - the cooperation was not constructive.

Both organizations started to question their decision-making structures, their relationship with each other and with other organizations.

7. Conclusions

The beginning of this case study was promising. The introduction of group-based PIM at village level was quite successful. In the light of failures and changes in the APP project, PIM expectations and indicators were subsequently changed, and the interest in observing and documenting was kept up. It finally switched from the project level to individual and group level, where changes and achievements are self-determined and less dependent on outside intervention. The farmers' group started to develop their own documentation sheets and ended up with a desire to learn and grow, and become more independent.

It takes time to internalize PIM and adapt it to one's own situation and the continuously changing environment. However, it is worthwhile investing time, because once PIM has been really understood it can be changed by the people themselves to fit into varying contexts.

PIM is a useful tool for self-help groups – to describe their own situation, to exercise their judgment and to clarify their goals vis–à–vis the NGO. For the NGO, too, PIM is useful for reflecting on goals, activities and instruments. But there are several obstacles to implementing PIM.

PIM is an instrument and a methodology. As an instrument, it helps people watch and analyze learning effects and all kinds of changes. As a methodology, it initiates learning and changes.

As PIM is a system for watching and analyzing reality jointly in a conscious and systematic way, it helps people to learn from project failures: the weakness of a project call for more independent solutions on the part of the farmers. Once identified as their own contribution, these actions will build up their self-esteem and their wish to continue to learn and grow.

PIM helps to bring out discrepancies between official project goals and farmers' real expectations and needs.

If the actors are strong enough to accept mistakes constructively without blaming each other, the prospects for learning from one another and for continuous change in the joint PIM workshops are good.

However, if the actors are weak and only try to defend themselves, conflicts will escalate and cooperation will stop. Handling problems and conflicts under such adverse circumstances will require additional know-how in conflict management and mediation.

If PIM is introduced in a highly hierarchical organization, more conflicts and obstacles arise. PIM requires flexibility and openness on the part of all parties involved. Moreover, PIM needs additional capacities and time for smooth cooperation to be ensured, especially in conflict-prone cases.

The observation of external and internal changes initiates further change processes. The latter are experienced more consciously by the people, and are therefore more persistent than changes induced from outside.

With SIBAT and PARTNERS, conflicts arose early: a project had been set up, and it responded too little to changing conditions and needs at grassroots level. Weakness led to fear of transparency, and ultimately to refusal to cooperate. There had been early signs of the problems which, much later, led to a reorganization of the project and project management. With PIM it was possible to record these early rumblings „seismographically".

In the long run, a monitoring system based on active participation of the grassroots is only feasible if the project really corresponds to people's needs. They will refuse to participate in monitoring for irrelevant activities and objectives. It seems that the existence of a functioning participatory M+E system is itself an indicator of people's interest in a project.

The general climate in development cooperation is often an obstacle to discussing problems openly. Fear of competition, fear of forfeiting professional respect and standing, and of losing projects and potential sources of funding - all these fears create an unproductive atmosphere, putting people on the defensive, making them play „hide-and-seek". This strategy is often adopted by all the actors involved - funding agencies (FA), NGO and groups - and complicates the business of establishing a constructive atmosphere of trust and confidence, mutual acceptance and learning.

Addresses:
SIBAT
P.O. Box 375
CPO Manila
Philippines

Dorsi Germann
FAKT – Association for Appropriate Technologies
Gänsheidestr. 43
D - 70184 Stuttgart, Germany

INDES in Misiones / Argentina

Promoting the development of a grassroots organization with PIM

*Resource Persons:
Cristian Krieger, Burkhard Schwarz*

1. Background

Development Organization / NGO

INDES is an NGO for rural development which operates in north-east Argentina. Its main purpose is to promote farmers' organizations and to support their projects with technical assistance and credits. INDES works in four provinces, with a relatively autonomous team of three to six staff members in each province. The INDES team in Misiones participated in the PIM field phase.

Self-help group

Unión y Progreso is a women's organization. It was set up in 1989 with 25 women. By 1994 membership had increased to 83. As the area covered by the organization is very large it is not easy for the members to meet. The group's main objective is to improve family living standards. Its principal projects are gardens to grow vegetables for family consumption, poultry production, training courses (in health and nutrition) and a credit scheme; the women hope to produce surpluses by selling their produce.

Formally, the organization is led by a management committee with 9 members, and has subgroups which nominate delegates. Informally, the president is a strong leader, supported by a nucleus of 5-6 committee members. The main concern of the members are tangible benefits, in return for which they offer political and religious loyalties.

2. Monitoring practice prior to introduction of PIM

Development organization / NGO

In the first phase of its existence, up to 1989, INDES made various attempts to institutionalize participatory methods of planning and evaluation. These were applied to some extent, in the form of

- monthly meetings between staff members and groups
- evaluation and planning workshops at area level every six months
- internal evaluation and planning workshops with staff members.

In the second phase, a financial crisis has led to a reduction in personnel numbers, greater autonomy of the teams, and activism. Periodical meetings with the groups which INDES supports continue, but no more internal workshops at INDES level have been held.

Self-help group

Unión y Progreso previously had no specific monitoring system. The group's management mechanisms are:

- fixed monthly meetings, with written minutes
- annual general meetings with presentation of the annual report and annual evaluation exercise (at the suggestion of INDES staff members)
- reflection meetings on specific problems (at the suggestion of INDES staff members)
- meetings for coordination and exchange of information with other groups and institutions.

At the last annual general meeting it was obvious that the group had considerable internal problems.

3. Introduction of the PIM concept

Development organization / NGO

At first INDES was reluctant to introduce PIM, due to a lack of internal discussion and because the field staff felt that the method would be too complicated for the group and INDES would not be able to maintain continuous close contact.

In 1993, the INDES team was strengthened by two young local promoters. On a ten-day visit a consultant helped to implement PIM. It was introduced at several working sessions and implemented in the following four steps:

- short conceptual discussion on participatory impact monitoring
- identification of organizational aspects as main monitoring topics
- reflection on expectations and fears of the INDES team
- definition of indicators and easy data collection tools.

The identification of organizational aspects as the main area to be monitored on behalf of INDES was an exercise for introduction of PIM at group level. Expectations and fears were reflected upon in the following five steps:

1. the principal issues were defined
2. expectations and fears were collected in a brain-storming session
3. expectations and fears were classified and synthesized
4. the synthesized items were transformed into indicators
5. the method of collecting data was determined.

Expectations

- that they know how to work
- that the proposed organization structure would lead to an increase in membership
- that growth of the group would have positive impacts on the social organization of the community and the appreciation of the women's role.
- that the capacity for collective management (of projects and administration) would increase
- that activism would still leave sufficient scope for reflection, planning and evaluation in the group.

Fears

- that the delegates would not be able to ensure satisfactory communication between the subgroups and the group as a whole
- that distortions in communication between the group and INDES would worsen
- that no operational leaderships would emerge which would dynamize the subgroups
- that power conflicts would emerge between delegates and management
- that roles and functions would become confused among the subgroups and the group
- that the increasing diversification of activities might exceed the group's management capacities and lead to failures
- that the organization of the group was directed too much towards material interests
- that the group's founders would reject new interests within the group that were different from their own.

Therefore, PIM was introduced mainly by applying it directly. The staff members faced some difficulties with the presentation of PIM; and due to a shortage of time, they did not have much opportunity to reflect on the concept.

Self-help group

The INDES staff's first attempts to introduce PIM in Unión y Progreso failed, due to a shortage of time in the group meetings and a lack of clarity and security among the staff. However, the fact that the subgroups did not work satisfactorily was a certain motivation to introduce monitoring mechanisms.

When the PIM advisor from Bolivia arrived, a special meeting was held between the INDES staff and women's group. This first meeting focussed exclusively on explaining how a monitoring system in general, and PIM in particular, works.

Two days later, there was a second meeting (3 hours) at which the members of the group agreed to apply PIM. Although they had not completely understood the monitoring concept, they agreed in the hope that monitoring would be useful for them and that they could learn by doing it. The women expressed their interest in improving, above all, their internal organization, especially the division into subgroups. Therefore, „participation and organization" was decided on as the main topic of monitoring. Occasionally, the group selected the same issue as the INDES staff; the staff were thus fully prepared to help establish an organization monitoring system for the group.

Next, the indicator system was developed in five steps:

1. the issue „participation and organization" was defined
2. expectations and fears concerning the women's own organization were collected in a brainstorming session
3. expectation and fears were grouped and synthesized
4. the synthesized items were transformed into indicators
5. the indicators were refined
6. the method of collecting data was determined.

There was not enough time to complete steps 4 to 6 for all the indicators at this same meeting. Finalization was therefore postponed, along with the selection of observers, until the next regular meeting.

A minor difficulty arose because in the brainstorming session the expectations and fears had not been documented properly, but only in catchwords. As a result, each catchword had to be rediscussed and interpreted. Some of the original ideas were thus lost.

4. Details of group-based impact monitoring

The indicators finally worked out by the members of *Unión y Progreso* included the following:

1. the number of organized subgroups has increased
2. the number of subgroups which meet in the course of the month has increased
3. the support provided by INDES staff in the activities of the subgroups has increased
4. the number of members who have surpluses of vegetables and eggs and who are selling them has increased.

It was difficult for the watchers to find time for their task. Even if they did not inform the group properly, they were nevertheless conscious of the indicators.

5. Details of NGO-based impact monitoring

The INDES staff defined 13 indicators which can be grouped in three categories:

1. operational aspects and internal communication
2. decision-making aspects
3. internal relations.

Examples of indicators:

1. number of group members
2. number of group members participating in monthly assemblies
3. number of group members participating in subgroup meetings
4. number of subgroups which send their delegates to the assemblies with mandates
5. number of autonomous activities of the subgroups
6. coordination of meetings was done by a member of the group.

The indicators were presented monthly in one of two ways:

– simply reading them, without graphs;
– visualization through graphs.

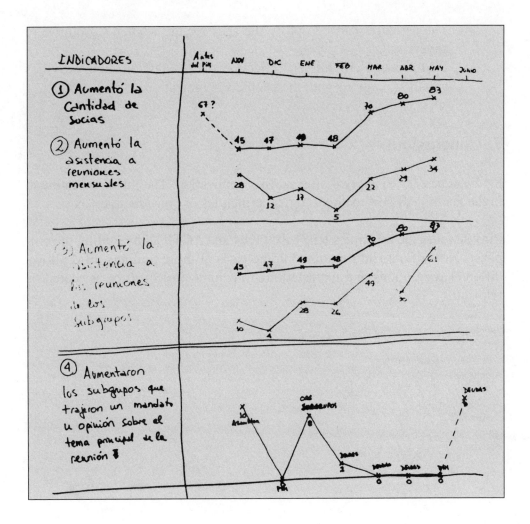

6. Impacts observed and induced by PIM

In the women's group, some of the impacts induced by PIM were observed immediately:

- The introduction of PIM has already generated decision-making.
- The formation of subgroups was accelerated and activities were transferred to this level.
- Roles and responsibilities were more strictly defined.
- The need to collect data for PIM led to a better internal communication structure.
- The presentation of INDES-based PIM prompted changes in responsibilities and in the decision-making structure.

There was also an impact within the INDES team:

- The implementation of PIM contributed to the formation of a field workers' team.
- The INDES staff had a better understanding of the group with which they worked.

- The activities carried out by the staff for the women's group were supervised.
- PIM induced critical self-reflection of the staff on their activities and their relations with the people's organization.

7. Conclusions

PIM can be easy to apply – after its first application. The initial investment, in the learning of new methodological principles, is considerable.

PIM enriches the principles REFLECTION and ACTION by putting the emphasis on verifiable information. On the basis of these findings, PIM allows different perceptions and expectations in a joint project to be accepted as relevant.

Addresses:
INDES Misiones
C.P. 3300, Posadas/Misiones
Argentina

Burkhard Schwarz
Casilla 395
Oruro
Bolivia

SUMMARY

What have we learned from the field studies?

This summarizing chapter is structured in the same way as the application examples. It attempts to sum up the experience gained so far and present it in a way that will enable future PIM users to avoid certain mistakes, or to shorten the learning process. Your application of PIM should produce good results right from the beginning, and not break the motivation of those involved before PIM has had a chance to unfold!

1. Background

> *This section emphasizes the preconditions and requirements for implementing PIM.*

Development organization / NGO

NGO motivation

The development organization or NGO should feel a need to introduce a transparent monitoring system. The personnel have to be motivated to apply PIM, because starting this monitoring concept will involve extra work. The additional inputs in cost and time, and how they are to be borne or financed, should be realistically clarified before PIM is introduced (SIBAT).

Participatory and intercultural approach

The NGO should have a participatory and intercultural approach to self-help support. Additionally, a stable organization structure and sufficient horizontal internal communication dynamics are desirable. If the information flow within the NGO is disrupted (merely due to structural instability or internal conflicts), the success of PIM will be limited (FEDECOMIN).

The NGO also has to be flexible enough to adapt and even change planned project activities if monitoring makes it clear that the initial plans do not correspond to the felt needs of the group (SIBAT).

Continuous information flow

For the application of PIM a continuous flow of information between the NGO and the people's organization (PO) is desirable; one aim of PIM is to improve this flow. If there are any major gaps to be bridged in communication, for example if the geographical distance is large, application of PIM will prove more difficult, but not impossible.

Self-help group

Autonomous and stable structure

The self-help group should have an autonomous structure. The NGO should not push the idea of implementing PIM too hard; rather, there should be a genuine interest in improving the running of a project. The PO must exhibit a minimum of organizational stability; there should not be any marked internal discrimination of members. Extreme migration dynamics make PIM more difficult to implement (Caracoles; SIBAT).

To a certain degree, horizontal formal or informal decision-making and communication structures should exist in the self-help group. If there are continuous disruptions of vertical power structures, the success of group-PIM as a guidance tool for the self-help group will be limited (Caracoles).

Further characteristics

Literacy of all members of the group is not necessarily a precondition for implementation of PIM (Caracoles).

Geographical concentration of members of the PO is an advantage. If the people participating in PIM are dispersed over a very large area, it is more difficult to apply (INDES). The same is true for organizational structures where adequate internal communication is hampered, e.g. by shift – working (Kantuta). This could be a serious obstacle to the sharing of PIM results which is vitally important.

Incidentally, all these factors also tend to determine the integral, vertical and horizontal multiplier effects relating to the PO's application of PIM.

2. Monitoring practice prior to introduction of PIM

This section attempts to identify the starting point for implementation of PIM, and discusses how existing elements can be used to build up PIM.

Development organization / NGO

Weak internal control mechanisms

In NGOs, monitoring is usually neglected due to a lack of time. Internal control and reporting systems are often lacking or not very effective.

NGOs' monitoring concepts tend to be rather vague if they exist at all. It is difficult for them to recognize or admit this, and consequently it seems difficult to clarify concepts in theory. The best way to start is to implement an actual monitoring procedure.

If a conventional monitoring system is already being applied it will reflect only the NGO's needs for project management, and not the views of the PO. Existing systematic monitoring practices are useful for the staff's understanding of PIM logics.

Communication between NGO and the self-help group

The NGO's knowledge of the group's internal structures is generally very limited. Usually there is no permanent flow of communication between the NGO and the self-help group. The usually unsystematic observation modes are exclusively based on the NGO's own initial knowledge of the group and the project context.

Self-help group

Control and change mechanisms

The groups have their own control mechanisms. Formally, there are regular meetings where basic decisions are taken, and elected leaders who put the decisions into practice. Informally, a group often depends on a few dominant persons who want to keep the power in their hands and not give away information.

There are almost always strategies for observing internal and external changes. They can sometimes be exploited for building PIM up organically, but normally they are not sufficiently systematic and frequently not participatory.

3. Introduction of the PIM concept

This section points out some critical aspects which were experienced in the field phase and which are prerequisites for successful introduction of PIM.

Development organization / NGO

Learning process

The introduction of PIM also involves a learning process for the NGO. The way in which the results of PIM are shared internally has to be clarified at the beginning.

It is advisable to introduce PIM gradually in one or two projects before introducing it on a large scale (PWDS). External advisors are often required when PIM is introduced for the first time, but they can be dispensed with if basic agreement has been reached concerning the implications of a participatory approach (PWDS).

Observation of socio-cultural issues

The socio-cultural issues should be developed in the context of concrete discussions about technical and economic issues. Special training in observing and assessing socio-cultural impacts and learning processes is useful. Staff members should give concrete examples based on personal experience.

Even for the staff it is helpful to develop PIM on the basis of traditional customs, in order to emphasize the importance of traditional structures and informal communication (FEDECOMIN).

Communication with the self-help group

The PIM design should consider the possibilities of information collection realistically, e.g. in the context of geographical distances separating the NGO from the self-help group, and communication difficulties.

It should be emphasized at the beginning that the NGO is helping the self-help group to introduce PIM. It is a kind of rehearsal if the introduction of PIM is begun by introducing first NGO-PIM and then group-based PIM; however, there is also a risk that the results of NGO-PIM will influence the results of group-based PIM.

Self-help group

PIM responsibilities and group structure

As mentioned above, the group's geographical and organizational structure should be considered in detail before defining the logics of PIM application. In particular, this is to avoid creating parallel political structures. It is recommended that the group's sub-structures be represented in the PIM observation team (FEDECOMIN).

PIM must be made compatible with the group's formal and informal meeting rhythm and logic. Although it is possible to work with PIM within the usual group meetings, the introduction of PIM requires a meeting of its own (INDES).

A conscious decision has to be taken as to whether those responsible for PIM should be linked to the group's decision-making level or separated from it. When choosing the PIM observation team, these alternatives have to be taken into account in the context of internal power structures. The „observation-reflection-action" cycle will be completed better if the self-help groups representatives participate in the PIM observation team (Kantuta).

Special training of the observers is indispensable, not only in order to create awareness but also for technical skills (e.g. measurement of pH) and illustration of the findings (PWDS, FEDECOMIN).

The introduction of PIM should be guided, if possible, by moderators who are „insiders". The development of observation methods should not be assigned to a committee; it should be done collectively (Caracoles).

People's culture and PIM

The PIM information base has to be built up by authentic insider knowledge of the reality of the group. The indicator system should be closely linked to the group's decisive and authentic priorities. The PIM design should also be adapted organically to expressions of the group's culture and its forms and logics of communication. Traditional ceremonies help to establish a basis of greater confidence between the group and NGO staff members. It is made clear that PIM is intended to be a flexible guidance tool which is compatible with the culture and the rules of the group (FEDECOMIN / SIBAT).

The PIM learning process will be delayed if the methodological approach does not correspond sufficiently to the cultural concepts of the group. In particular, the way observations are shared within the group has to be clarified.

Identification of socio-cultural impacts

The socio-cultural issues should be developed around practical discussions on technical and economic issues. The case studies of PWDS and FEDECOMIN have shown clearly that when the protagonists identify indicators for changes that are subjectively important to them (because they will change their lives considerably) they utilize very tangible „technical-economic" illustrations which often stand for socio-cultural impacts.

Not every expectation or fear is related to the impact level. Although this may lead to theoretical discussion, in practice it is no problem because operational or short-term expectations/fears, if converted to indicators, will soon be replaced by the group, whereas the continuous expectations/fears will be maintained.

Visualization

Pictures are crucially important as aids to explaining PIM. The visualized message stimulates learning processes through both cognitive and emotional responses. Humour and enjoyment help to reduce stress, boredom and resistance (SIBAT).

However, although visualization has proved useful, it is not necessarily always the culturally appropriate method (PWDS). Especially in indigenous contexts, the oral tradition should be considered for new ways of applying PIM. Visualization should not be given sole preference.

4. Details of group-based impact monitoring

> *This section illustrates some typical handicaps of group-based PIM during the first trials, and how they can be overcome.*

General framework

First, it should be emphasized that during the field phase all the self-help groups succeeded in setting up and running a group-based monitoring system. However, the degrees of intensiveness and ways in which it was done were context-specific. The application of PIM was quite irregular at the beginning unless the NGO carried out intensive follow-up.

The more political support PIM enjoys within the group, the more successfully it will be applied and the more useful it will be. Members of the group must understand the sense of comparing observed results as a fundamental part of the system. Continuity in the PIM observation team, i.e. observation by the same individuals over a prolonged period of time, is crucially important.

If decision-making cannot be influenced by monitoring, people lose their motivation to apply it (SIBAT).

Indicators

It is useful to formulate complete questions to make the meaning of an indicator clear. This is also helpful in remembering the meaning at subsequent meetings, and in encouraging active contributions by members of the group (FEDECOMIN).

There is no need to work out scientific indicators. Tables and graphs have been used for all numeric indicators. Although they will be new to some members of the group at the beginning of the monitoring exercises, people become familiar with them and lose their fear of statistics. Even the YES/NO indicators have proved meaningful if they are complemented with the question „Why?" or „Further observations?" (FEDECOMIN).

The monitoring of group performance as the only issue dealt with by PIM is possible but does not seem to be self-sustained in the long run (SIBAT, INDES).

One handicap in the implementation of PIM may be that highly confidential or „sensitive" internal information will not be transformed into indicators by the group, although this would be the most important issue. Structural difficulties of this type can only be overcome by more internal and autonomous management of PIM, i.e. by the group itself.

Joint reflection workshops

In the joint reflection workshops also, PIM should be guided by moderators who are also insiders. However, when PIM is first implemented, group-PIM has to be facilitated by members of the NGO's staff. It is for this reason that, at first, there is sometimes no clear distinction between group-based PIM and joint reflection workshops - but it should develop in the long run. The PIM team from FEDECOMIN decided to hold joint reflection meetings once a month, and PWDS even twice a month, by introducing PIM elements in their regular meetings.

Joint reflection meetings between members of the NGO's staff and the group are held more frequently in the initial stages of PIM. In the long run they will probably be superseded by joint reflection „workshops" which are held only once or twice a year to analyze the project impacts in greater depth.

At the joint reflection meetings/workshops it is helpful to present the results of group-based impact monitoring first and those of NGO-based impact monitoring later. The comparative nature of PIM can be emphasized by visualization or equivalent methodological approaches.

The dynamics of change should lead to smooth adaptation of PIM tools in the joint reflection meetings/workshops. Continued observation of factors which have already been dealt with and are obsolete makes PIM boring (Kantuta).

If there are serious difficulties in completing the „observation-reflection-action" cycle, it may be useful to monitor the efficiency of decision-making and of adjustments induced by PIM as well. Even the efficiency with which PIM results are shared internally could be assessed.

Joint reporting by various groups or people's organizations which carry out similar projects makes good sense. This exchange of experience induces direct learning processes among the people themselves, as the groups can learn better from each other (PWDS).

Adjustments

The members of the group should try to assess their experience with PIM independently. To prevent PIM drifting away from reality, the group must develop adjustments of PIM themselves, corresponding to the increase in their familiarity with the problem.

5. Details of NGO-based impact monitoring

> *In this section, similarly to the previous one, methods of overcoming some handicaps in the implementation of NGO-PIM are described.*

General framework

As for the self-help group, continuity of the PIM observation committee, i.e. observation by the same individuals for a prolonged period of time, is also crucial for the NGO. The staff must also fully understand the logics of comparison of the observed results as a fundamental part of the PIM monitoring system.

Frequent rapid assessment of the application of PIM, considering both NGO and group level, is useful for adapting it to the dynamics of the environment (FEDECOMIN). Intercultural problems between NGO and the group and the logics of normal communication flow between the two sides should be taken into account. Internal sharing of PIM results, and reflection on the efficiency of the method of sharing internally, should not be forgotten.

If there are serious difficulties in completing the „observation-reflection-action" cycle, it may be useful to monitor the efficiency of decision-making and of adjustment of the results of PIM observation even in NGO-based impact monitoring.

Indicators

Although NGO-based impact monitoring mainly focuses on socio-cultural impacts, the technical and economic issues are likewise monitored. As we have seen above, they are often an expression of expectations and fears relating to socio-cultural learning processes. Moreover, they have a considerable influence on the success of the project if no conventional monitoring system was previously used (FEDECOMIN, PWDS).

There is a risk that too many indicators will be worked out, focussing on similar aspects. Such monitoring tends to be boring. It will not be done regularly or with the necessary drive and may be useless within a short period of time (INDES).

Unexpected impacts

One of the main tasks of the NGO staff during the joint reflection meeting/workshop is to resolve the following methodological problem: while expected or supposed impacts have been identified, unexpected impacts may be more important. They are surprising, unpredictable events and as such are not directly covered or taken into account when the indicators are defined.

This leads to the difficulty that unexpected impacts have to be recorded separately. The best way to do this seems to be to reserve some time at joint meetings exclusively for this purpose.

Joint reflection workshops

Good preparation and management of the joint reflection workshops and meetings is crucial to the success of PIM implementation. Both workshops and meetings should be jointly moderated. The NGO staff member should take the chair alone only during that part which concerns the NGO; s/he should not dominate the entire meeting. Indicators should be recorded by each side before the joint reflection workshop/meeting, not during the workshop itself.

Generally, visualization seems to be the most suitable and culturally adapted tool for NGO level during the joint reflection meeting.

6. Impacts observed and induced by PIM

This section attempts to illustrate the link between subjective and objective impact assessment.

General remarks

PIM reveals the complexity of people's expectations and of project impacts. PIM goes beyond project issues, it clarifies the project context and the situation of the group.

The impacts observed with PIM seem less important than the impacts induced by PIM.

Impacts on communication between self-help groups and NGO

PIM helps to consolidate actual economic projects (PWDS, Kantuta, Caracoles), and to develop new management capacities (Kantuta). This is mainly due to the fact that PIM can help to overcome problems of intercultural communication between NGO and groups.

Conscious observation of changes and the joint reflection meetings improve communication between the NGO and the self-help groups, and this leads to collective learning processes. PIM also produces, on the one hand, NGO knowledge about groups and about itself, and on the other hand, group knowledge about NGO and about itself.

Impacts on participation, organization and decision making structure

PIM generates self-confidence, in particular at group level (Caracoles, SIBAT). It creates awareness of the responsibilities of the parties involved in the project, and helps the parties assume those responsibilities. The permanent recording of changes facilitates decision-making. Thus, PIM offers an opportunity to adjust plans and adapt the project to the needs of the people, or for staff members to initiate corrective action.

The group has an opportunity to define the main issues of its project activities. The NGO's staff have an opportunity to find out what the group thinks and feels about the project. PIM reveals that it is often different from what staffs expect, or that key implications of a project have not been understood. Also, within an organization, PIM can lead to the creation of new communication instances (Kantuta) and even change the organization's structure.

PIM makes the organization transparent, and tends to change the de facto decision-making structure. For this reason it also provokes resistance from those in power. This includes a potential danger that can lead to self-destruction of PIM.

7. Recommendations for avoiding basic risks

> *The PIM field phase has elicited many critical reflections on this monitoring concept. On the basis of the internal evaluation the following recommendations, relating to certain basic risks encountered during the field phase, have been formulated:*

- The key factors in successful implementation of PIM are the continuity of its application, the sharing of its results and a proper communication structure.
- PIM necessarily has to take the specific cultural normative systems of the actors as the basis for estimates and expectations concerning its impact
- PIM should not lead to increasing interventions by NGOs in the self-help group's affairs, and thus to the reduction of the group's autonomy in management (especially during the joint reflection meetings). A mutual respect for different levels of confidential and „sensitive" information management is the basis for the efficiency of PIM.
- People's reality and at the same time their knowledge of and familiarity with problems are highly dynamic. Accordingly, the imple-

mentation of PIM has to be adapted dynamically to theses changes.
- PIM is intended to lead rapidly to an alleviation or a substantial reduction in workloads. This means that the methods necessarily have to be simple, and their application has to be quick and easy.
- During the application of PIM, there should always exist absolute clarity about the separation or conjunction of the two levels decision-making on one hand and PIM observation on the other. It will influence the selection of personnel and the contents of the monitoring.
- PIM can promote internal emancipation organically only if there is a genuine political consensus on this subject. If not, PIM may provoke more internal contradictions, and this can lead to its self-destruction.

8. Conclusion

PIM is not only a method for monitoring projects. It is at the same time an instrument for planning and for the development of teams and organizations and it induces individual learning processes for all parties involved.

✽

Although PIM is simple, it does require a certain amount of effort to teach its contents. The first organizations in which PIM was introduced did not understand the concept entirely. Here the introduction was accompanied by more intensive support. In one organization, however, where only a basic introduction to PIM was given, the PIM concept was quickly implemented. The successful communication of the concept of PIM still needs to be clarified.

PIM is a young concept which needs to be developed further. Right from its beginning it has been concieved together with partners from all over the world – in a process-orientated way.

We need the cooperation of other practitioners and thinkers to test and to improve PIM.

If you are implementing and testing PIM in your project area we would be very interested to hear from you. Write and tell us about your experience with PIM.

We are planning to organize more regular and more efficient exchanges, if a substantial number of practitioners continue with the development and adaptation of PIM.

Please write to

FAKT	or	GTZ - GATE (ISAT)
Association for Appropriate Technologies		German Appropriate Technology Exchange
Gänsheidestraße 43		Postfach 5180
D - 70184 Stuttgart, Germany		D - 65726 Eschborn, Germany

Thank You